CONTESTING AFRICA'S NEW GREEN REVOLUTION

Politics and Development in Contemporary Africa

Published by one of the world's leading publishers on African issues, "Politics and Development in Contemporary Africa" seeks to provide accessible but in-depth analysis of key contemporary issues affecting countries within the continent. Featuring a wealth of empirical material and case study detail, and focusing on a diverse range of subject matter—from conflict to gender, development to the environment—the series is a platform for scholars to present original and often provocative arguments. Selected titles in the series are published in association with the International African Institute.

The principal aim of the International African Institute is to promote scholarly understanding of Africa, notably its changing societies, cultures, and languages. Founded in 1926 and based in London, it supports a range of seminars and publications, including the journal *Africa*.
www.internationalafricaninstitute.org

Managing Editor: Max Vickers
Series Editors: Jon Schubert (Brunel University) and Elliot Green (London School of Economics and Political Science)

Editorial Board

Already published:

CONTESTING AFRICA'S NEW GREEN REVOLUTION

Biotechnology and Philanthrocapitalist Development in Ghana

Jacqueline A. Ignatova

ZED

Zed Books
Bloomsbury Publishing Plc
50 Bedford Square, London, WC1B 3DP, UK
1385 Broadway, New York, NY 10018, USA
29 Earlsfort Terrace, Dublin 2, Ireland

BLOOMSBURY and Zed Books are trademarks of Bloomsbury Publishing Plc

First published in Great Britain 2021
This paperback edition published in 2023

Series design by Burgess & Beech
Cover Image: A member of the Rural Women's Farmers Association of Ghana RUWFAG,
Near Lawra, Ghana. (© Global Justice Now)

ISBN: HB: 978-1-7869-9655-8
PB: 978-1-7869-9656-5
ePDF: 978-1-7869-9657-2
eBook: 978-1-7869-9658-9

Series: Politics and Development in Contemporary Africa

Typeset by Newgen KnowledgeWorks Pvt. Ltd., Chennai, India

To find out more about our authors and books visit www.bloomsbury.com
and sign up for our newsletters.

www.bloomsbury.com

CONTENTS

AUTHOR BIOGRAPHY

Jacqueline A. Ignatova is an Assistant Professor of Sustainable Development at Appalachian State University in Boone, North Carolina. Her work has been featured in *Third World Quarterly* and *African and Black Diaspora: An International Journal.*

PREFACE

When I was preparing my application for a Fulbright fellowship in 2011 to study the unfolding debate around genetically modified crops in Ghana, I knew that I might be a bit "early to the scene." I got reassurance by my workshop participation and interviews in DC that Ghana would soon be both a new site for confined field trials of genetically modified crops and a target of "new Green Revolution for Africa" interventions promoted by a global assemblage of actors in development, agriculture, and philanthropy. "Ghana is the furthest along," with regard to the introduction of genetically modified crops in West Africa, one DC-based development official knowledgeable about global biotech trends told me.[1] Yet, when it came to identifying a Ghanaian anti-GMO or food sovereignty–related activist organization tracking these developments, I came up short.

I decided to send an email to Bakari Sadiq Nyari, someone I understood from my office in Baltimore to be a food and land sovereignty activist, who had written a compelling exposé of biofuel land grabbing in northern Ghana. He had passionately condemned a Norwegian biofuel company that tried to establish "the largest jatropha[2] plantation in the world" for violating the land rights of a community: "Bypassing official development authorization and using methods that hark back to the darkest days of colonialism, this investor claimed legal ownership of these lands by deceiving an illiterate chief to sign away 38,000 hectares with his thumb print" (Nyari 2008: 1). By highlighting this "land grab," Nyari's work led to a return of the land to the community and has been used as an example of success by transnational activists in the struggle against land grabs across Africa.[3] A few months following my e-mail, Mr. Nyari wrote that he would be happy to work with me and would connect me to the executive director of RAINS, a Tamale-based NGO that focuses on intergenerational knowledge exchange in agriculture and gender equity work in education. If there was not yet a big debate about GMOs in northern Ghana, I thought an organization like RAINS and Nyari's work on land grabbing would be the likely source of its murmurings.

It turned out that Mr. Nyari is far more than a food and land sovereignty activist. Nyari—contrary to my initial impression upon reading his exposé—is a "big man" in Ghana and through his former service as national chairman of Ghana's Lands Commission has been involved in part of these so-called new Green Revolution (nGR) interventions. Additionally, Nyari was the Director of the Lands Commission's Public and Vested Lands Management Division and the Principal Lands Officer of the Upper East Regional Lands Commission, as well as a part-time university lecturer in land economics. He co-owns a hotel with his wife, serves as a principal consultant at Premium Property Consult, has an extensive garden, raises quail, and is growing the Alliance for a Green Revolution in Africa

(AGRA)'s jasmine rice varietal. In his capacity as chairman of RAINS, he has supported their work on seed sovereignty and gender equity, and has repeatedly stressed its importance: "women play a key role in making this choice" of what seed to plant, and so a loss of control over the seed system would disproportionately affect women.[4] In Nyari's words, "whoever controls the seed, controls the farmer, and whoever controls the farmer, controls the food."[5]

Nyari believes that the "communal nature of land" needs to be protected and has used his positions of authority to do so, as I discuss further in the interlude, "On 'Mixing.'" As Nyari told me in one of our interviews, "Our land really holds us together and if we destroy its communal nature, we have also destroyed the unity that exists between us."[6] These views on the need to retain the communal nature of customary land tenure sit in tension with projects such as the World Bank–funded Ghana Commercial Agriculture Project (GCAP) that advance land's commodification. However, Nyari's influence in the design of the GCAP land bank—mixing lived concerns about land grabs with his expertise in land economics—has led to the specification that only a percentage of the land could be sold in the form of limited time leases, appearing to strike a balance between the realities of rural-to-urban migration and high land values, without denying communities access to land.

What gets lost in thinking of Nyari as merely a Ghanaian bureaucrat that implements decisions made by external funders—as some conventional accounts of the new Green Revolution in Africa insinuate—is his diverse concerns and multiple role performances: expert, entrepreneur, activist, educator, community leader. Claims by both proponents and opponents of a new Green Revolution in Africa (and its related introduction of genetically modified crops) as wholly transformative obscure the fact that Ghanaian elites have multiple and competing loyalties, that African farmers are pragmatic, and that the work of "development" always remains unfinished. Rather than a "revolution" from or "termination" of traditional farming practices, pragmatic "people's science" guides farming decisions as farmers stay open to new, promising opportunities to resolve the challenges that they face and to improve and diversify their livelihoods (Richards 1985; Berry 1993; Scoones 2009, 2015; Scott 1998: 264). Instead of generating a clear trajectory of outcomes, new Green Revolution interventions are likely to be "mixed" with existing traditional knowledge and practices and modified by a range of competing livelihood concerns.

Since 2011 my fieldwork in Ghana has enabled me to follow an evolving conversation about these nGR interventions. My analysis traces the debate around these interventions from a time when few people discussed GMOs and global agribusiness in Ghana, particularly in northern Ghana (2011–13), to a period when activists marched against Monsanto in the streets of Accra and educated farmers and students reversed the role of interview subject and *asked me* for my views on genetically modified crops, unprompted (2015–present). I have witnessed shifts in the responses of US development planners from dodging my questions (or restricting their answers to off-the-record) about their long-term views on the transformation of the Ghanaian seed sector, to explicitly stating the goal and role of US interests in these new Green Revolution for Africa interventions. This book is a study of a period of transition, and what I have learned is the product of fieldwork approached with an ethnographic sensibility, engaging in participant observation and repeated conversations and interviews with a diverse array of key figures involved in Ghanaian agricultural politics between 2011 and 2018. The blend of fieldwork and discourse analysis has informed my core argument: the "new Green Revolution for Africa" is not really new, not really "green," not a revolution, and not "pro-poor"—as is implied by its designation that it is "for" Africa. Allow me to take each of these points in turn.

It's not really new, but presenting it as such has important political effects. A look at agricultural policies in Africa in the 1960s and 1970s shows that Green Revolution programs didn't miss Africa as proponents of the new Green Revolution for Africa claim; rather, they were either not sustained (Djurfeldt et al. 2005; Patel 2013; Wiemers 2015; Goldman and Smith 1995; Akram-Lodhi 2013: 84) or ill-suited for the African context (Berry 1993; Richards 1985). However, as I argue in Chapter 1, there are political benefits of framing these efforts to "transform" African agricultural systems as "new" and "uniquely"[7] promising, given how such a discursive framing helps to obscure the role that some of the nGR's biggest proponents have played in creating the conditions for its supposed necessity. These claims of newness also illustrate the deployment of hype to stimulate technological adoption and philanthropic support, which I discuss in Chapter 4.

Take for example nGR agricultural interventions on rice in Ghana. Rice is the subject of genetic manipulation in the development of a nitrogen and water use-efficient salt tolerant (NEWEST) rice at the Crops Research Institute in Ghana, a purportedly pro-poor biotechnology crop that is the product of the "donation" of "trait technologies" from Arcadia Biosciences (AATF n.d.d). (I describe a parallel example of philanthrocapitalism[8] with the development of a transgenic cowpea in Chapter 3.) Rice is one of the three crops promoted as part of USAID's Feed the Future programming and was a key crop of the Ghana Commercial Agriculture Project. Although the development of biotechnology is relatively new, it is driven by many of the same actors and the same practices of extraction of plant germplasm for the development of new technologies we have seen in the first Green Revolution, on which I elaborate in Chapters 1 and 3. The World Bank and US interests lead the charge in promoting rice productivity now; however, these actors were among the most influential in promoting structural adjustment programs (SAPs) that resulted in the decline of the Ghanaian rice industry. Trade liberalization enabled Ghanaian markets to be flooded with heavily subsidized imported rice that forced a huge proportion of Ghanaian rice farmers to cease rice production due to an inability to compete (Mittal 2009). Furthermore, as I discuss in Chapter 2, public–private partnerships are needed now to address weaknesses in agricultural research and development as well as extension capacity due to SAP-related cuts in agricultural spending—from 12.23 percent of total government expenditures in 1980 to 0.39 percent in 2007 (Chambers et al. 2014: 27). The means and message on agricultural development spending in Africa may have shifted over time, yet it is new configurations of the same players that deliver "authoritative" development advice.

It's not really "green." A widespread misinterpretation of these agricultural interventions as "green"—environmentally sound and sustainable—was first brought to my attention when I gave a lecture at a Ghanaian university in 2013. I told the event organizers about my research and gave the title of the talk: "Seeds of contestation: The 'new Green Revolution' in Ghana." Despite the title and the description of the talk that emphasized my critique of initiatives to modernize African agriculture and the use of the language of "hotly contested technological models," the flyer image used had a strikingly different tone. It showed a rendering of green paper doll cutouts with hands held to encompass planet Earth. Out of the top of the earth was a single sprout of a seed. In contrast to this common trope of sustainability, the nGR promotes monoculture and a narrow focus on seeds as technology and a source of new market value (biocapital). As I discuss in the introduction, monocultural practices reduce diversity and rely more on fossil fuel–derived inputs and mechanization. The potential expansion of a commercial seed sector is likely to reduce the varietals planted and consequentially biodiversity—thereby also diminishing an important source of resilience in the context of climate change and, ironically, the source of new commercial traits. Additionally, the wisdom of promoting water-intensive crops such as rice is questionable given the challenges of erratic rainfall and dried-up dams faced by smallholder farmers, discussed further in Chapter 5. While proponents claim that nGR interventions promote "sustainable intensification" that reduces the need to expand cultivated areas, the productivist logic (McKeon 2015) that drives these interventions neglects issues of distribution, waste, and the significant dependence on fossil fuels that sustain global agricultural value chains.

It's not a revolution. If we take a revolution to mean a sudden, rapid, and complete change, neither the first Green Revolution nor it's "new" version meets that criteria. Rather, as my interlude on what I call "mixers" highlights, farmers make pragmatic decisions based on what is available within their means and shaped by different forms of knowledge. As such, they are not likely to embrace totalizing technological reform. Although this talk of "revolution" and "transformation" dominates conversations among foreign development planners and elites of Ghana, it lacks popular circulation, particularly among farming communities in northern Ghana. The agricultural development interventions denoted as part of this nGR require *sustained and considerable* capital investment, training, and knowledge production and dissemination in order to recruit farmers.

A particular trope within this nGR discourse, used frequently in the Ghanaian context, is that these nGR interventions will carry Ghana "from farming as a way of life to farming as a business." This suggests that there is an abandonment of the previous spatiotemporal position—farming as a way of life is replaced by a business mentality—but also a radical break from existing forms of knowledge, property, and agency, as I discuss in the introduction and the interlude. Yet, just like the case of Nyari's multiple, coexistent roles as expert/entrepreneur/activist, on the ground there is quite a bit of friction between these two ideal types. Model farmers chosen to lead some of the nGR's most prominent interventions in Ghana continue to bring to bear traditional farming wisdom, communal notions of property, and distributive notions of agency[9] alongside the use of these new agricultural technologies and registered land titles. Rather than being supplanted, these practices are mixed and coexist—hardly a revolution. This is likely not a surprising finding to anyone who has spent time in the field studying agricultural development or has read Sara Berry or Paul Richards or Ian Scoones or James Scott. Nevertheless, this discourse of revolution persists and, as such, needs to be contested.

Which leads me to my final point—*it's not "pro-poor."* An anecdote from one of the first contacts I made is illustrative here. I had met this professional development assessor through a professor of mine, a former research assistant from a community on the outskirts of Tamale whose expertise was in monitoring and evaluation. He had worked with a number of NGOs and government agencies on a contract basis. During our first meeting in January 2012, I remarked at the surprisingly high number of NGOs in Tamale—many of which were focused on agriculture and food security—and asked him how locals felt about this NGO presence. Did they feel this improved upon their lives? Did it feel invasive? He told me he thought that "more people should study this as these are interventions into peoples' lives and they don't always make peoples' lives better … sometimes they make them worse." He cited an example of a community in the Upper East that had numerous development interventions and no improvement. "Why is that?" I asked. He said that the aid "would not always reach the target beneficiaries … sometimes they select students that are already brilliant and would succeed regardless to show success." What his story suggests is that, again, agricultural development interventions to promote productivity and food security are not new and that much development assistance continues to reach people that have the means to be successful regardless—rather than choosing beneficiaries that are most in need. Chambers's (1983) argument reminds us that this tendency in rural development is not a new phenomenon. Philanthrocapitalism (described in the introduction and in Chapter 2) influences many nGR interventions and exacerbates this already existing tendency in agricultural development to deepen processes of social differentiation, as discussed in Chapter 5.

This consultant's story also reveals some of the paradoxes of humanitarian food security work. He was, at the time, the single person hired to do the monitoring and evaluation work for the northern Ghanaian programs of an international NGO. Despite his workload, his employment was paid substantially less than that of his Euro-American colleagues (some of whom were also provided housing) on short-term contracts. At the time we spoke of it, he was already six months into a year-long contract without knowledge of whether he would be renewed, and he worried about what this would mean for his ability to support and feed his large family. The irony of his employment in the domain of food security was not lost on either of us, and learning about his insecure status attuned me to view global food security and development interventions with a more skeptical eye.

The presentation of these high-modernist agricultural development interventions as a pro-poor new Green Revolution for Africa is not only inaccurate but also deceptive. It obscures the political reality that these interventions sustain a status quo that has been harmful to African development: one that has created the conditions of dependence on Western expertise even while contradictory and damaging, extracts African knowledge of the use of plants for profit even while demeaning traditional African farming practices, and uses the cover of pro-poor philanthropy to obscure how it also creates the conditions for the development of global agribusiness that cares little about resource-poor African farmers. If the new Green Revolution for Africa is really none of these things, *what are these interventions about? Who is really developing whom?* This book seeks answers to these questions.

ACKNOWLEDGMENTS

It is hard to properly thank all of the people who helped shape this decade-long project. I want to first thank my friends and informants in Ghana who constantly challenged my thinking. In particular, I am grateful for the friendship and conversations over the years with Bakari Sadiq Nyari, Edward Salifu Mahama, Mamudu Akudugu and his students, Faiza and Kaka Taimako, Stephen Mahama, Helen and David Azupogo, Edwin Andoh Baffour, Bern Guri, Eric Okoree, Lansah Alhassan, Issahaku Alhassan, Thomas Fuseini, Daniel Olad, Miles Adongo, Christopher Azaare, Bonaba, Ishmael Salifu, Spooner Atamale, Sheila Salifu, Uncle Musa, and Feo Naba who have shaped my thinking about poverty and agricultural development in Ghana. I am grateful to the people of Kpegu Bugurugu, Kukuo Yapalsi, Bongo, Vea, Nyariga, and Feo for welcoming me. I am also grateful to the community I met through the US Embassy during my Fulbright (grant #34122686), which facilitated connections to key actors involved in agricultural development. I am particularly thankful for the generosity and support of the Banashek family, Stephen Perry, Fara Jim Awindor, and Nii Sarpei. I also extend tremendous gratitude to Brad Horwitz, who made my pilot trip to Ghana possible in 2011. Abel Atimbire, Eddie Annan, and Sam not only helped me get to where I needed to go, but the conversations during these rides are ones I look back upon fondly. I am also extremely grateful to Mary, Titi, Louisa, Evelyn, and Mperba for their friendship and sharing their compound with me.

This book's foundation is my doctoral dissertation, *Seeds of Contestation*, completed within the Department of Government and Politics at the University of Maryland (UMD), College Park, in 2015. I thank my mentor Ken Conca for consistently asking tough questions that helped me to develop my ideas and for encouraging me to do fieldwork; he may well appreciate how many of the great books we read together during an independent study at UMD have now featured prominently in this book. I thank Virginia Haufler for asking me to consider whether some of the phenomena I had identified as new really were, as well as for offering her insights into the role of the private sector in contemporary politics. Additionally, I am thankful to have been able to work with Karol Soltan, John McCauley, Isabella Alcañiz, Patricio Korzeniewicz, Mary Kate Schneider, Mark Shirk, Daniel Owens, Jennifer Wallace, Michael Beevers, Rodrigo Pinto, Laryssa Chomiak, Marty Kobren, and Jonathan Hensley at UMD, who provided me with feedback on earlier versions of this project. I also thank the UMD Program for Society and the Environment as well as the Department of Government and Politics for travel and intellectual support.

My experience living in Baltimore and bearing witness to structural racism was impactful and further encouraged me to pursue activist scholarship. When my partner, Anatoli Ignatov, was accepted to Johns Hopkins, Jane Bennett encouraged the two of us to live in Baltimore and made a promise that the community of political theorists at Hopkins would welcome me. Such promises were more than realized, and my connection with the Hopkins Political Science Department and African Seminar helped me grow tremendously as a scholar. I am especially grateful to Sara Berry for feedback on both earlier versions of chapters and the book proposal as well as for including me in the African Seminar. Her generosity taught me to be a different kind of scholar. The opportunity to share my early fieldwork observations in a room full of brilliant Africanists pushed me to use my fieldwork more effectively and to be more attentive to historical processes that shape contemporary Ghanaian politics. I am particularly grateful for the feedback on fieldwork and theory from Casey McNeill, Alice Wiemers, Lori Leonard, Siba Grovogui, Jeff Ahlman, Julia Cummiskey, Jane Guyer, Pier Larson, Kevan Harris, Jane Bennett, Bill Connolly, and Chad Shomura. Mimi Keck's research and writing workshop was pivotal in developing my work on the political economy of hype and in encouraging me to write in an accessible style that included stories; I am also grateful for the feedback from Beth Mendenhall, Devin Fernandes, and Tarek Tutunji during these sessions.

This book has also benefited from my participation in the International Studies Association annual meetings over the years. I am particularly grateful for the conversations with and feedback on earlier versions of chapters from Jennifer Clapp, Simon Nicholson, Nora McKeon, Michael Spann, Matias E. Margulis, Matthew Eagleton-Pierce, Noah Zerbe, Sarah Martin, Jennifer Lawrence, Brian Dowd-Uribe, Allen Stack, Shiera Malik, Jacob Stump, Kevin Funk, Olivia Umurerwa Rutazibwa, and Cecelia Lynch. My understandings about philanthrocapitalism and public–private partnerships benefited from conversations at the African Studies Association Annual Convention with Carol Thompson, Rachel Schurman, and William Munro. I am also indebted to Marcus Taylor for inviting me to be a participant in the Dimensions of Political Ecology conference and have benefited from his work and feedback as well as that of Andrew Flachs, Glenn Davis Stone, and Garrett Graddy-Lovelace.

A major turning point for me as a scholar was my participation in the Institute for Qualitative and Multi-Method Research in Syracuse and the opportunity to meet other scholars interested in interpretive methods and fieldwork. In particular, the boundary-pushing work of Timothy Pachirat, Lisa Wedeen, Victoria Hattam, Tanya Schwarz, Biko Koenig, Nick Smith, Iris van Huis, Brian Alan Guy, Daragh Grant, Stephen Cauchon, Maren Bjune, Paloma Raggo, Merouan Mekouar, and Maite Tapia has continued to inspire me. An earlier version of Chapter 3 was published in *Third World Quarterly*, and I am grateful for the feedback of two anonymous reviewers that helped me sharpen my argument as to what is at stake with these political economic transformations of the "new Green Revolution" in Africa. I am also thankful for feedback from anonymous reviewers on two article submissions to *Global Environmental Politics* and *Journal of Agrarian Change* that challenged me and pushed me to be more explicit about my commitments to interpretivist social science.

The book proposal benefited from the feedback and advice of Ken Barlow, Stefanie Fishel, Jessica Martell, Desiree Fields, Kathleen Krull, Mauro J. Caraccioli, Dana Powell, Jennifer Westerman, Kira Jumet, and Isaac Kamola that demystified the book writing process and gave me critical editorial suggestions. My fantastic interdisciplinary colleagues at Appalachian State University that include Brian Burke, Rebecca Witter, Aniseh Bro, Rick Rheingans, Christof den Biggelaar, and Valerie Wieskamp gave me excellent critiques of earlier versions of chapters that made this work more capable of reaching a wide audience. Research assistants Emily Hubbard, Kellen Mahoney, and Devyn Barron enriched this project by sharing with me which themes grabbed their attention and what areas needed better clarification; Jamie Hedrick provided critical assistance on the index. Getting to welcome James Scott to Appalachian State was a life achievement unlocked, and when I had the opportunity to share with him my book project, Jim asked me why I was using the language of the "Green Revolution," remarking: "it's not green and it's not a revolution." This conversation has inspired the preface and book title that places the contestation of these terms front and center. I am indebted to the incredibly helpful organizational and clarifying suggestions for the full manuscript to two thoughtful anonymous reviewers and the editorial team of Zed/Bloomsbury as well as excellent editorial suggestions on my introduction from Elizabeth Bennett.

Lastly, my family's consistent encouragement has been a priceless form of support. In particular, I would like to thank my best friend, husband, travel companion, and editor extraordinaire Anatoli Ignatov, who has been so supportive throughout this process. Anatoli was an active participant in the African Seminar, IQMR, and Appalachian State University writing groups and has been with me in my journey every step of the way. Our fieldwork trips to Ghana together were always the source of such joy, and my project would not be the same had we not had day-long interviews together with Chris Azaare covering topics that exceeded my own research agenda but enriched it beyond what words can express. The countless hours of conversations and ideas exchanged made this project what it is today. Анатоли, обичам те с цялото си сърце.

ABBREVIATIONS

3ADI	African Accelerated Agribusiness and Agro-Industries Development Initiative
AATF	African Agricultural Technology Foundation
ABS	access and benefit sharing
ACDI/VOCA	Agricultural Cooperative Development International/Volunteers in Overseas Cooperative Assistance
ADVANCE	Agricultural Development and Value Chain Enhancement
AGOA	African Growth and Opportunity Act
AGRA	Alliance for a Green Revolution in Africa
AMPLIFIES	Assist in the Management of Poultry and Layer Industries with Feed Improvements and Efficiency Strategies
ARC	Institute for Agricultural Research (South Africa)
ATT	Agriculture Technology Transfer
AU	African Union
BMGF	Bill & Melinda Gates Foundation
BNARI	Biotechnology and Nuclear Agriculture Research Institute
Bt	*Bacillus thuringiensis*
CAADP	Comprehensive Africa Agriculture Development Programme
CABI	Centre for Agriculture and Bioscience International
CEO	chief executive officer
CFT	confined field trial
CGIAR	Consultative Group on International Agricultural Research
CIAT	International Center for Tropical Agriculture
CIKOD	Centre for Indigenous Knowledge and Organizational Development
CIMMYT	International Maize and Wheat Improvement Center
CIP	International Potato Center
COFAM	Coalition for Farmers' Rights and Advocacy Against GMOs
CPP	Convention People's Party
CRI	Crops Research Institute (Ghana)
CSIR	Council for Scientific and Industrial Research (Ghana)
CSIRO	Commonwealth Scientific and Industrial Research Organisation (Australia)
DFID	Department for International Development (UK)
ECOWAS	Economic Community of West African States
FARA	Forum for Agricultural Research in Africa
FAO	United Nations Food and Agriculture Organization
FAS	Foreign Agricultural Service (US)
FAW	fall armyworm
FDI	foreign direct investment

FinGAP	Financing Ghanaian Agriculture Project
FSG	Food Sovereignty Ghana
FTF	Feed the Future
G7	Group of Seven
G8	Group of Eight
GAIN	Global Agricultural Information Network
GAPs	good agricultural practices
GCAP	Ghana Commercial Agriculture Project
GE	genetically engineered
GIIN	Global Impact Investing Network
GM	genetically modified
GMO	genetically modified organism
GNAFF	Ghana National Association of Farmers and Fishermen
GR	"Green Revolution"
IAR	Institute for Agricultural Research (Nigeria)
ICRISAT	International Crops Research Institute for Semi-Arid Tropics
ICT	information and communication technologies
IFAD	International Fund for Agricultural Development
IFDC	International Fertilizer Development Center
IFPRI	International Food Policy Research Institute
IITA	International Institute of Tropical Agriculture
IMF	International Monetary Fund
INERA	Institute of Environment and Agricultural Research (Burkina Faso)
IP	intellectual property
IPRs	intellectual property rights
IRRI	International Rice Research Institute
ISAAA	International Service for the Acquisition of Agri-Biotech Applications
JSR	joint sector review
KARI	Kenya Agricultural Research Institute
KNUST	Kwame Nkrumah University of Science and Technology
MDGs	United Nations Millennium Development Goals
MOAP	Market-Oriented Agricultural Programme
MoFA	Ministry of Food and Agriculture (Ghana)
MPRRI	Madhya Pradesh Rice Research Institute
MVP	Millennium Villages Project
NEPAD	New Partnership for Africa's Development
NEWEST	Nitrogen and Water Use-Efficient Salt Tolerant Rice Project
NGO	nongovernmental organization
nGR	"new Green Revolution"
OFAB	Open Forum for Agricultural Biotechnology in Africa
PASS	Program for Africa's Seed Systems
PBB	Plant Breeders' Bill
PBS	Program for Biosafety Systems
PPP	public–private partnership
RAINS	Regional Advisory Information and Network Systems (Ghana)
SADA	Savannah Accelerated Development Authority

SAP	structural adjustment program
SARI	Savannah Agricultural Research Institute (Ghana)
TRIPS	Trade-Related Aspects of Intellectual Property Rights
UNDP	United Nations Development Programme
UNESOC	United Nations Economic and Social Council
UNIDO	United Nations Industrial Development Organization
UPOV	Union for Plant Variety Protection
USAID	United States Agency for International Development
USDA	United States Department of Agriculture
VAD	vitamin A deficiency
WACCI	West African Centre for Crop Improvement
WB	World Bank
WEF	World Economic Forum
WEMA	Water Efficient Maize for Africa
WISHH	World Initiative for Soy in Human Health
WTO	World Trade Organization

INTRODUCTION

In 2011, Ghanaian president John Atta Mills signed into law Parliament's Biosafety Act 831, which allows for the cultivation of genetically modified (GM) crops in Ghana.[1] This important piece of legislation followed the 2010 Ghana Plants and Fertilizer Act,[2] which allows for the privatization of seed production and created the initial changes to "release some of those [seed] regulatory policies to the private sector."[3] Both bills passed without much public discussion—but not without the work of the United States Agency for International Development (USAID)'s Program for Biosafety Systems, which hosted workshops for parliamentarians years prior in order to "facilitate sensible and dispassionate debate" regarding biotechnology[4] (USDA as quoted in Rock 2018: 118).

More than three years after the Biosafety Act passed, activists from Food Sovereignty Ghana sued the Ghana National Biosafety Committee and the Ministry of Food and Agriculture for failure to abide by the provisions of the Biosafety Act, as a means to ban the commercialization of genetically modified organisms (GMOs).[5] The Ghana National Association of Farmers and Fishermen (GNAFF)—an organization eager to share with me their experience touring the Monsanto-founded Danforth Institute of Plant Science[6]—joined the case in defense of the introduction of GM crops. The political party of Kwame Nkrumah, the Convention People's Party (CPP), joined the plaintiffs in order to reject the "imposition" of GMOs; GMOs represented the "neocolonization" that the first president of Ghana had warned about in his writings.[7] The court case is generating a growing public debate about GMOs in Ghana today, prompting questions regarding who has voice in agricultural decision-making, how contested agrarian futures are mediated, and who, ultimately, has the power to set development priorities. As one member of Food Sovereignty Ghana told me in the context of the court case, "The world is watching Ghana."[8]

Ghana's agricultural future is currently at a critical juncture. The introduction of GM and hybrid seeds as part of the "new Green Revolution" of "agricultural transformation" has been cast, on the one hand, as the technological savior to address Africa's food insecurity (e.g., Paarlberg 2008; Water Efficient Maize for Africa Project) and, on the other, as the Trojan horse of corporate neocolonialism (e.g., Daño 2007; African Centre for Biodiversity's[9] campaign against GMOs in South Africa). The debate is further intensified by aid agencies, foundations, and media voices that portray Africa as a deficient continent with a "starving" population in "urgent" need of a "new Green Revolution for Africa."[10] Proponents of GM seeds, like the Bill & Melinda Gates Foundation and USAID, endorse biotechnology as a means to improve crop yields and adapt to changing environmental conditions.[11] Opponents like La Via Campesina and the African Biodiversity Network cast this new technology as threatening to cultural and biological diversity and smallholder farmers' self-sufficiency. Ghana's developmental success in West Africa has made the country an important site for these development interventions, with experts anticipating that if Ghana chooses to adopt GM seeds and increase the commercialization of its agriculture, others will likely follow suit.

Population growth, the effects of climate change, and the combined shift in interest of global finance and biotechnology have placed African agriculture on the international agenda. Global venture capital has turned its attention to the African continent as a place of enormous potential, ripe for investment: "Africa is the final frontier—the last sizable area of untapped growth in the global economy. To succeed, companies will need to bring Africa into the boardroom" (Dupoix et al. 2014: n.p.).[12] A 2014 World Economic Forum Annual Meeting echoed this sentiment by pointing out that six of the top ten fastest growing economies were in Africa: Ethiopia, the Democratic Republic of the Congo, Côte d'Ivoire, Mozambique, Tanzania, and Rwanda. Africa's abundance of natural resources, 60 percent uncultivated arable land, and the largest global workforce were highlighted as other factors to excite investment on the continent (World Economic Forum 2014; Holodny 2015). Investment in African agriculture is perceived to be more profitable now for a number of reasons: at a time of global economic downturn and a global shortage of arable land, the African savannah is a frontier for new capital investment and emerging markets; the financialization of agriculture has made the trade in commodity futures possible and lucrative; legislative change within African countries has created a more secure "enabling environment" for investors; and technological change and increased opportunities for the use of information and communication technologies (ICT) in agriculture have increased expectations that agriculture can be profitable in these market frontiers.[13]

One of the key manifestations of this shift in attention toward African agricultural systems is the emergence of powerful actors and partnerships pushing for a new Green Revolution for Africa. The "Green Revolution" is a referent to the high crop yields that countries in Asia and Latin America experienced in the mid-twentieth century through the introduction of scientific seeds, agrochemicals, irrigation, and linkage to markets. This "new Green Revolution" (nGR) is advanced by a global assemblage of actors that includes governments, such as the United States and Ghana; international and bilateral aid agencies, such as the World Bank and USAID; foundations, such as the Bill & Melinda Gates Foundation and the Rockefeller Foundation; international and national agricultural research institutions, such as CGIAR (Consultative Group on International Agricultural Research) and the Savannah Agricultural Research Institute; and agribusiness, such as Monsanto[14] and Syngenta. Institutions such as the Alliance for a Green Revolution in Africa, "born of a strategic partnership between the Bill & Melinda Gates Foundation and the Rockefeller Foundation to dramatically improve

African agriculture, and to do so as rapidly as possible," bring together public and private resources (AGRA n.d.e.). The work advocated and executed by these actors is intended to address poverty and perceived flaws in African farming systems through a "productivity revolution in smallholder farming" (World Bank 2007: 1).

I argue that the instrument effects (Ferguson 1994; Foucault 1979) of the new Green Revolution for Africa agenda are something quite different than these "pro-poor" aspirations. This assemblage of philanthropists, development experts, investors, and global agribusiness has shown considerable commitment to the commercialization of seeds and the related increased privatization of plant breeding, regulation, agricultural extension, and research on the African continent. Utilizing multisited fieldwork in Ghana between December 2011 and August 2018 that examines these efforts to promote a commercial seed sector and the entry of GM crops into Ghanaian markets, I aim to shed light on the agenda of the new Green Revolution for Africa and problematize its claims that it is "new," "green," "revolutionary," or "pro-poor." In contrast to its claims as "African solutions to African problems," the nGR instead privileges particular forms of knowledge and agency in the production of agrarian capital that reproduce global power inequities and advance social differentiation in Ghana. As I discuss in the preface, it is critical to ask *in whose interests* this agenda of agricultural transformation is pursued.

In order to set the foundations for the book's central query, I explore three themes: knowledge, capital, and agency. I then discuss my methodological approach and describe the chapters ahead.

Knowledge, Development, and the "Logic of Extroversion"

The willful disdain for local competence shown by most agricultural specialists was not, I believe, simply a case of prejudice (of the educated, urban, and Westernized elite toward the peasantry). … Rather, official attitudes were also a matter of institutional privilege. To the degree that the cultivators' practices were presumed reasonable until proven otherwise, to the degree that specialists might learn as much from the farmer as vice versa, and to the degree that specialists had to negotiate with farmers as political equals, would the basic premise behind the officials' institutional status and power be undermined. (Scott 1998: 286)

James Scott reminds us that scientific knowledge can, and often does, suppress local forms of knowledge due to its privileged (albeit insecure) position within modern institutions. As Helen Tilly (2011: 70) describes in *Africa as a Living Laboratory: Empire, Development, and the Problem of Scientific Knowledge, 1870–1950*, the production and dissemination of knowledge formed an effective governance strategy for imperial rule on the African continent. Scientific knowledge supported by technical experts funded by the British Empire came to dominate development decisions, displacing vernacular knowledge (ibid.). Knowledge and expertise itself was an expression of colonial rule, and the institutionalization of scientific knowledge helped to ensure its perpetuity. Scientific knowledge was both developed and applied in colonial Africa; Africans became active agents in the appropriation, production, and application of scientific knowledge (Tilly 2011: 14).

The work of experts that could claim to know Africa and would support "development" on the continent became a crucial element in the rationalization of a continuation of empire in the post–Second World War period (Cooper 2002). Colonial knowledge of Africa developed through years of research and social engineering enabled the British officials to acquire an authority grounded in expertise on African affairs. Although scientists tended to separate ecological and agronomic topics (e.g., crop varieties, diseases, cultivation methods) from farmers' knowledge in order to study the former, colonial officials' "knowledge" of Africa derived largely from their experience in trying to get taxes collected, roads cleared, crops sent to market, and so forth, without provoking too much dissent.[15] This privileging of Western expertise and interests has only intensified with structural adjustment programs that altered the Ghanaian food economy (where food demands were increasingly met by imports, rather than domestic production) and substantially reduced employment in the government and agricultural research sectors, accelerating the turn toward the commercialization of African farming systems that the World Bank's "agriculture for development" agenda implores (World Bank 2007).

Another way to sustain these unequal power relations is to naturalize them and to obscure from view the political. Timothy Mitchell (2002) describes the ways in which development planners have sought to disengage themselves from the political realities that they have helped shape. In his book on the politics of technical expertise in twentieth-century Egypt, *The Rule of Experts*, Mitchell looks at the ways in which problems in the developing world have been presented by development planners as problems of mismanagement, knowledge, or nature, rather than political issues. Such a framing of the problem as one that is technical or managerial gives experts political authority to determine the solutions—what Tania Li refers to as "rendering technical" (Mitchell 2002; Li 2007). In a similar vein, James Ferguson (1994: 256) identifies that although development projects in Lesotho have power effects, the development industry never allows its role to be formulated as a political one. Development intervention in Lesotho premised on the "promise of agricultural transformation appears simply as a point of entry for an intervention of a very different character," that is, the reinforcement and expansion of the exercise of bureaucratic state power alongside the depoliticization of both poverty and the state (Ferguson 1994: 255–6).

Similarly, *Contesting Africa's New Green Revolution* shows how numerous agricultural development programs in Ghana today reflect the interpretation of agrarian challenges as mere technical, managerial issues. These programs focus on agricultural technology transfer and training, the formalization of intellectual property and plant breeders' rights into law, and the expansion of university programs in genetics, plant breeding, and "seed science and technology"—supported by universities such as the University of Illinois at Urbana-Champaign, Cornell University, and Iowa State University. The depoliticization of agricultural development interventions is aided by this work of agricultural "global learning" initiatives that link institutions of higher education, which I discuss in Chapter 6. The power of scientific knowledge is reflected in the ways in which farmers' experimentation is labeled in a derogatory fashion as haphazard "trial and error," whereas plant biotechnology is presented by industry as offering precise means of achieving agricultural advancement.

In reality, both farmers' experimentation and plant biotechnology research and development involve trial and error, albeit at different temporal rhythms (Shiva 2000).[16] Funding priorities to support agricultural biotechnology—rather than resource-efficient traditional agricultural practices—are another indication of these power relations. As discussed in Chapter 4, the legacy of the "rule of experts" is further reflected in the very terms of the debate over Africa's agricultural future—where it is being held (in the public sphere or in exclusionary places like courts of law, as is the case regarding GMOs in Ghana), what knowledge is needed to be taken seriously (such as the ability to invoke science and legal expertise), who is given voice, when, and to what effect.

Paulin Hountondji (1997b: 12) characterizes these dynamics as "the logic of extroversion" "whereby we expect motivations, initiatives and starting signals for our deeds to originate from places other than our own societies." Applying the logic of extroversion to Third World countries furthermore neglects

> the debt European science owes to the Third World, the nature and the scope of knowledge resulting from the theoretical processing of the voluminous quantities of fresh data extracted from there, the real function of the new disciplines based on that knowledge (tropical geography, tropical agriculture, African sociology, anthropology, etc.) and the shifts and shake-outs caused in old disciplines by these new discoveries. (Hountondji 1997b: 4)

Hountondji points out that there has been far greater attention paid to the material dimensions of extraction that contribute to African underdevelopment; attention to such intellectual extraction is warranted given how it "enhanced the slow but inexorable integration of the entire proto-scientific and scientific legacy of the South … into a process of intellectual production on a world scale managed and controlled by the North" (1997b: 5, 7).

In contrast to this logic of extroversion, endogenous development "is based mainly, though not exclusively, on locally available resources, local knowledge, culture and leadership, and their cosmovisions, with the openness to integrate outside knowledge and practices" (Haverkort, Millar, and Gonese 2003: 6) and "aims at the local determination of development options: local control[17] over the development process and the retention of the benefits of development within the local area" (Millar 2014: 640). It further resonates with Richards's (1985: 162) endorsement of a "people's science," "a peasant-focused, decentralized approach to research and development in West African agriculture," which recognizes farmers as experimenters and innovators capable of producing "good science." Both endogenous development and people's science respond to this logic of extroversion by celebrating local knowledge, resources, experimentations and other practices as ways for communities to solve their own problems. These ideas run parallel to the idea of food sovereignty that I discuss in Chapter 6.

The book's concern with the "friction" (Tsing 2005) of these contesting knowledge systems builds on the work of Kloppenburg (2004), Shiva (1991, 1997, 2000), Yapa (1996a), Scott (1998), and Stone (2007, 2010), all of which examine the socioeconomic and political-ecological impacts of agricultural technologies. As I discuss in Chapter 5, efforts to promote a shift in Ghana from "farming as a way of life" to "farming as a business" align with Green Revolution interventions that "devalued the 'reproductive power' of nature by substituting the 'productive power' of industrial inputs" (Yapa 1996a: 82). This presumption of substitutability is what Shiva describes as a "mechanistic"—as opposed to ecological—viewpoint that claims to be objective, "value-free," but is nevertheless "compatible with the needs of commercial capitalism" (Shiva 1991: 23). Shiva goes further, asserting that "the Green Revolution was based on the assumption that technology is a superior substitute for nature, and hence a means of producing growth, unconstrained by nature's limits" (Shiva 1991: 15). She points to the violence of Green Revolution technologies that, although framed as ways to transcend the scarcities of nature, created "new scarcities in nature through ecological destruction" as well as conflict-inducing social differentiation (ibid.). Further violence is produced when the wisdom of previous generations of farmers is devalued, local seed saving and exchange is disrupted, rituals to support good harvests are no longer practiced, and labor is no longer communally shared.

One of the great forms of violence of the Green Revolution and an expression of the logic of extroversion has been the expansion of monocultures. When I speak of *monoculture* I consider six elements that are a part of the new Green Revolution for Africa interventions and highlighted throughout Chapter 5: (1) knowledge that informs agricultural practices, (2) modality of farming, (3) farmer identity, (4) type of seeds, (5) conception of property,[18] and (6) market structure. Efforts to promote a new Green Revolution for Africa, and in particular to commercialize the seed sector in Ghana, can advance monoculture and the rule of experts at the expense of regenerative agriculture and food sovereignty (further discussed in the section on agency). I will take these points in turn.

What Shiva (1997) terms a "monoculture of the mind" is an outcome of training, production, and dissemination of knowledge about "good agricultural practices,"[19] the use of demonstration farms, and the expansion of Western-influenced formal education and educational exchanges. Monoculture as a modality of farming is visible in the logic of agricultural exit, an idea articulated in the World Bank's influential *Agriculture for Development* report, which invites those farmers that are not a part of the "productivity revolution in smallholder agriculture" to leave farming (2007: 1). This particular modality of farming is further reinforced through the dissemination of "starter packets" of seed and inputs as well as participation in "value chains" that standardize and integrate production, processing, and distribution. A particular kind of farmer identity—that of the entrepreneurial "serious" farmer—is promoted in agricultural workshops, farmer trainings, field visits, tours, and videos played in remote villages intending to reach "last mile users" of agricultural technologies. The impoverishment of biodiversity emerges as a consequence of this standardization inherent to productivism (McKeon 2015; Shiva 1997), whereby a limited range of crop varietals are mass produced as cash crops to meet the demands of markets near and far. Western understandings of property shape the formalization of land and seed that are an integral part of nGR interventions. These conceptions of property clash with the diversity of existing community property relations and practices of seed exchange (Bratspies 2007; Mgbeoji 2006; Shiva 2000). Likewise, the market structure monoculture that emphasizes value chains and outgrower schemes shapes labor, exchange, and food distribution in ways that may undermine the conditions of possibility for polycultures based in labor pooling, seed saving, and traditional knowledge.

Monocultures are highly legible but ecologically vulnerable systems (Scott 1998). Scott defines legibility as the simplification, standardization, and formalization of nature and space in order to support state and economy making. Industrial agricultural systems—due to mechanization, standardization, the cultivation of monocultures, and the formulaic use of inputs (pesticides, herbicides, fertilizers)—are highly legible systems designed to maximize agricultural productivity. Because of this narrowed focus on agricultural productivity, these systems have also had to face chemical-resistant "superweeds" and "pests," soil nutrient depletion, and reduced biodiversity—consequences that emerge from outside of this constricted field of vision. Despite both the extensive efforts that go into the standardization of agricultural practices and its undesirable ecological effects, such work is depoliticized by its supporters. Moreover, the emergence of these negative externalities are frequently interpreted as problems of knowledge, management, and technology that only additional expertise can resolve.

Capital, Bioeconomy, and "Corporate Geopolitics"

The appeal of rendering complex agricultural systems legible is that it facilitates capital accumulation; corporate consolidation in the seed, aggregation, and agrochemical sectors reveals that monocultures have been tremendously profitable for global agribusiness (Clapp 2012a; Clapp and Fuchs 2009; Kloppenburg 2004; Millstone and Lang 2008). The successful shift toward commercial agriculture on the African continent has been presented as contingent upon the formalization of land rights and seed law "harmonization,"[20] trends that align with the global movement to standardize intellectual property rights as embodied within the World Trade Organization's Trade-Related Aspects of Intellectual Property Rights (TRIPS) Agreement. Through legislative changes that "harmonize" differences and align with global agreements, the complex plurality of land tenure systems and seed systems is rendered legible, facilitating land transactions and the growth of a commercial seed sector that entices investment and commercialization of the agricultural sector. In Ghana, and many other African countries, seed production had in the past been associated with national food security and the public sector.

Legibility is created both through the deployment of tunnel vision to simplify complex reality (Scott 1998: 11) and by way of the cartographer's view that creates the "object of development" through a distancing view from above (Mitchell 2002: 209). Both techniques—the constriction of vision and the cartographer's view—radically simplify the object of focus, making it easier for the state or capital to act upon it. I argue that one of the goals of the work of the new Green Revolution assemblage is to render more legible agricultural and legal systems in order to create entry points for the market expansion of global agribusiness into African markets, particularly within the seed sector. My fieldwork in Ghana provides an illustration of what these dynamics look like.

However, before I advance this argument, a bit more reflection is needed on the processes that enable this capital accumulation and the emergence of "corporate agrifood governance" (Clapp and Fuchs 2009: 1), whereby corporations play a key role in influencing the rules that regulate their own behavior. In the paragraphs that follow, I discuss briefly the process of enclosure and its expression within the contemporary politics of seed in Ghana. The latter half of this section looks at the emergence of two new forms of capital—*philanthrocapital* and *biocapital*—that are features of what I call the "corporate geopolitics" of the new Green Revolution for Africa agenda. It situates such analysis within the domain of biopolitics.

The term "enclosure" has its historical roots in the enclosing of common pastureland that began in sixteenth-century England. Historian J. M. Neeson (1993: 15) identifies enclosure as the "extinction of common right" that led to the decline of small farms and the creation of a class of landless laborers. The legal process of enclosure allowed landholders to acquire other parcels of land, which created large farms and enabled landholders to claim private ownership. By the eighteenth century, the law itself had become a mechanism of the expropriation of common land from the peasantry (Marx [1867] 1976: 885). Enclosure is part of a set of practices that facilitated "the great transformation"—the formation of the market society in modern Europe—whereby land and labor are disembedded from their social and ecological context to facilitate their commodification (Polanyi [1944] 2001: 71–80).

The concept of enclosure can be used to describe the politics of contemporary acquisition of fundamental bases for life processes—land and seed—through dispossession (White, Borras, Hall, Scoones, and Wolford 2012). On the African continent, there has been growing concern[21] about the practice of land grabbing, the "exploration, negotiations, acquisitions or leasing, settlement and exploitation of the land resource, specifically to attain energy and food security through export to investors' countries and other markets" (Matondi, Havnevik, and Beyene 2011: 1; Nyari 2008). New land enclosures are the subject of considerable scholarly attention (e.g., White et al. 2012; Matondi, Havnevik, and Beyene 2012; Borras, Hall, Scoones, White, and Wolford 2011; Li 2011; Nyantakyi-Frimpong and Bezner Kerr 2017; La Via Campesina 2012; Fairhead, Leach, and Scoones 2012; McMichael 2012; Manji 2006; Cotula 2013; Kaag and Zoomers 2014a), but it is seed enclosures that are of particular concern to this book.

These new enclosures of land and seed are what David Harvey calls "accumulation by dispossession," a term he substitutes for Marx's concept of primitive accumulation in order to highlight the ongoing process of accumulation. Such accumulation originates from the moment in which agricultural producers are severed from their means of subsistence: the land and, now more recently, the seed (Marx [1867] 1976; Shiva 2000; Fitting 2011). The enclosure of seed occurs through legal regimes such as the WTO's TRIPS Agreement that allows the patenting of seeds. This legal maneuver is used to protect GM seeds, now understood by law as an information technology (US Committee on Science House of Representatives 2003: 65; Manu 2016). Patents on seeds are significant for four reasons, discussed further in Chapter 3: first, it is now possible to assert private ownership over life in the form of a seed; second, private ownership may disrupt the traditional practice of seed saving that has served as an important practice of resilience; third, this proprietary regime disregards the local and traditional knowledge of the use of plants that have been a foundation for plant biotechnology and other research developments; and fourth, there is no compensation for prior development of the seeds.

In his examination of the historical origins of the political economy of plant biotechnology, Jack Kloppenburg (2004) details how germplasm[22] derived from the Third World was appropriated for crop improvement in the "gene-poor" industrialized world at little cost and required no compensation to these "gene-rich" sources. This transfer of genetic material institutionalized by CGIAR used indigenous landraces[23] as raw materials for the subsequent breeding of hybrid high-yield varieties used during the first Green Revolution (Kloppenburg 2004: 161). The patenting of transgenic seeds based off of generations of farmers' experimentation raises important questions about the valorization of knowledge and the meaning of intellectual property, that is, whose knowledge is privileged and protected, as well as why and how.

The development of proprietary seeds—seeds protected by a patent—has raised alarms regarding the ways in which farmers' experimentation and knowledge can be used to develop seeds that cannot be freely saved or shared later. This form of enclosure has been deemed by some as "biopiracy," that is, "the unauthorized commercial use of biological resources and/or associated traditional knowledge, or the patenting of spurious inventions based on such knowledge, without compensation" (Mgbeoji 2006: 13). In the Ghanaian context, enclosure is found in the privatization of plant breeding, the encroachment of Western-style intellectual property protections, and regimes of biosafety and "seed harmonization," discussed in Chapters 3 and 6. My fieldwork reveals that this process of enclosure—the extraction of the target of enclosure (seed, land) from its larger ecological and social context thereby enabling its commodification—has been both driven by, and is in the interests of, a corporate geopolitics.

By "corporate geopolitics" I mean the ways in which global corporations are able to open up new markets and spheres of influence by way of state power. This form of enclosure is characteristic of what scholars of the global food economy refer to as the "corporate food regime"[24] or "corporate agrifood governance" (McMichael

2005; Clapp and Fuchs 2009). That is to say, corporations, by working closely with the public sector, have helped to shape the rules governing trade, intellectual property, and food production, particularly within the World Trade Organization (Korten 2001). Drawing upon the insights of Karl Polanyi ([1944] 2001) on the "self-regulating market" as a set of instituted processes, as well as Susan Strange (1996) and Michel Foucault (2008) on the role of the state in regulating *for* (rather than *because of*) the market, this book examines how the state performs an enabling role to promote corporate interests, both in the United States and in Ghana. Through many conversations during fieldwork in Ghana between 2011 and 2018, I found that US State Department, US Department of Agriculture, and USAID officials generally understood the promotion of agricultural biotechnology and the interests of American businesses as part of their responsibilities. Some of these actors collapsed the welfare of the poultry, beef, and soy industry or companies such as Monsanto with the national interests of the United States. Furthermore, the US State Department's Office of Agricultural Policy reflects the preeminence of American business interests in foreign policy with the ultimate goal to "suppor[t] American agriculture while protecting U.S. national security" (US Department of State n.d.).

I argue that a corporate geopolitics is expressed through the discourse of a new Green Revolution for Africa. Although it presents itself as humanitarian work and a novel movement of technology transfer, this discourse is entangled with the advancement of corporate interests eager to expand into new markets to increase their profits. Whereas, as discussed in Chapter 1, both phases of the Green Revolution have been geopolitical in nature, the novelty of the new Green Revolution for Africa is found in the advent of *philanthrocapital* and *biocapital*, two related forms of capital that have emerged as key new sources of market value. Philanthrocapital and biocapital work in conjunction with one another to rationalize the commodification of seeds through the promise of humanitarian benefit. An effect of this privatization of seed and knowledge commons is the further extraction of knowledge from African countries, even while demeaning African knowledge and practices and advancing Western notions of property.

The emergency discourse of population pressures, climate change, and technological deficiency that is characteristic of Green Revolution discourses (discussed in Chapter 4) charges African countries to do more to address food insecurity. This discursive frame promises a salvationary technological revolution that can turn things around without getting into the messiness of structural change, land tenure reform, or redress of gender, class, and ethnic inequities. Philanthrocapital—the merging of the logic of venture capitalism with philanthropy—assures that humanitarian and business interests can coexist and be mutually beneficial (e.g., Bishop and Green 2009). However, this "win-win" logic says nothing about the distribution and scale of such wins. Development influenced by the logics of philanthrocapitalism, what I term "philanthrocapitalist development," views investment in entrepreneurs and "high impact" areas as an adequate poverty alleviation strategy—despite the thin empirics of the "trickle-down" economic approach. Chapter 2 discusses further the influence of philanthrocapitalism on development and builds on the work of Thompson (2014), McGoey (2015, 2014), and Morvaridi (2012).

In turn, I situate my discussion of biocapital—the capitalization of life itself and a novel outcome of the new Green Revolution for Africa interventions—within the literature on the bioeconomy and biopolitics, which highlights how the life sciences and "enterprising nature" have become key areas of capital expansion and sites of power struggles (e.g., Sunder Rajan 2006, 2012; Jasanoff 2012; Cooper 2008; Helmreich 2008; Shiva 1997; Birch and Tyfield 2013; Birch 2017; Dempsey 2016; Foucault 1979; Brooks 2005; Rose 2001). Melissa Cooper (2008: 4) calls for greater attention to this domain of study: "Now, more than ever before, we need to be responsive to the intense traffic between the biological and the economic spheres." The representation of seed in informational terms, as intellectual property and a commodity that can be bought and sold, is a way by which *life itself* has been turned into a business plan (Sunder Rajan 2006: 41). As discussed in Chapter 3, biocapital is created through the expropriation of traditional and indigenous knowledge, publicly funded research, and years of international collaboration to enable the privatization and commodification of seed commons. Biocapital is that fullest expression of enclosure: it not only privatizes that which was for the "common heritage" of humankind, but it also shifts understandings of how life itself (i.e., seeds, transgenes, DNA sequences, human tissue) can be reduced to both intellectual property and a commodity.

In order for biocapital to be reproduced, legal systems are necessary to protect this newly defined property and the rights of the "innovator." Michel Foucault's *The Birth of Biopolitics: Lectures at the Collège de France, 1978–1979* is illuminating here. Foucault notes that with the growth of an "enterprise society," a key characteristic of the neoliberal state, this multiplication of enterprises increases the need for legal arbitration (Foucault 2008: 149–50). The promotion of entrepreneurship—an expression of the "enterprise society"—is at the core of the new Green Revolution for Africa agenda and philanthrocapitalist development. Given Foucault's insights, it is unsurprising that legislative reforms of African states to create an enabling environment for investment have been at the center of this agricultural transformation agenda, as "a society orientated towards the enterprise and a society framed by a multiplicity of judicial institutions, are two faces of a single phenomenon" (Foucault 2008: 150). I discuss the enabling environment for investment in the following chapter and the promotion of entrepreneurship in Chapter 5.

Foucault states that *Homo œconomicus*—whom he views as synonymous with the entrepreneur[25]—is at once an individual that pursues his own rational self-interest and must be left alone (*laissez-faire*), but also is the

person who accepts reality and responds systematically to modifications in the variables of the environment, appears precisely as someone manageable, someone who responds systematically to systematic modifications artificially introduced into the environment. *Homo œconomicus* is eminently governable. (Foucault 2008: 270)

Taking these insights into account suggests that the new Green Revolution for Africa agenda not only alters power relations by creating space for a rule of experts and litigators but also enhances the governability of populations through the promotion of entrepreneurship.

In other words, new Green Revolution interventions promote new forms of capital accumulation that are at once geopolitical and biopolitical. Raj Patel (2013) argues that this new Green Revolution is rather part of "the Long Green Revolution" that is both continuous and dynamic. Patel states that "one of the most striking features of the calls for a 'New Green Revolution' is the deployment of terms like 'nutrition' in novel ways" and analyzes this shift by engaging with Foucault's notion of biopolitics (Patel 2013: 4). African Green Revolution interventions that promote nutrition—many of which are funded by the Bill & Melinda Gates Foundation, Howard G. Buffett Foundation, and USAID—emphasize school supplemental feeding programs, genetic engineering of micronutrient-enhanced crops, as well as the general monitoring and evaluation of nutrition as part of the promotion of public health. This emphasis on nutrition as part of this new Green Revolution agenda neglects how "the Green Revolution has undermined human nutrition ... through its displacing of nutritionally rich food crops with commodity crops. Traditional micro-nutrient rich crops such as pulses, vegetables and fruits have been substituted for cereal grains which have a much lower nutritional value" (Patel 2013: 47). Patel links the legacy of the Green Revolution to distortions in the global food system that have left global populations both "stuffed and starved" (referring to the 1 billion people suffering from undernourishment alongside the 1.5 billion people overweight on empty calories (Patel 2012: 1; 2013: 47)). This inattention to the harmful dietary shifts that the Green Revolution helped usher in is one example of how development actors neglect their own role in creating the conditions for their intervention, a theme further developed in Chapter 1. The relationship of the biopolitics of nutrition and its connection to the governance of humanitarianism is explored next.

Agency and the Governance of Humanitarianism

These biopolitical interventions related to biotechnology and nutrition form part of what Michael Barnett describes as the "governance of humanitarianism" and raise profound questions about agency. In general, the development of biotechnology is still supply-led, rather than demand-driven, as it is still dependent upon the transfer of proprietary technology from multinational corporations (discussed further in Chapter 3). In numerous examples in the development of pro-poor biotechnology, the targeted beneficiaries—particularly women farmers—were inadequately consulted, or not at all (e.g., Zerbe 2005; Glover 2010b; Ignatova 2017). Food aid is another example of a nutrition-related intervention that reflects this governance of humanitarianism: while meeting caloric requirements, food aid frequently falls short on cultural appropriateness,[26] often reflecting the interest of donor countries to dispose of surplus grain (Clapp

2012b). This lack of consultation of recipient populations in humanitarian work is not uncommon, as Barnett (2011: 221) explicates:

> It is only over the last few decades that humanitarian governance has incorporated the views of local populations—and it is debatable how much energy humanitarians have put into these efforts or how receptive they are to redirection. … Humanitarians offer many reasons why it is difficult to adhere to modern standards of participation, but the implication is that the legitimacy of humanitarian governance does not depend on a process of deliberation, dialogue, or even consent.

I suggest that we can see these dynamics at play in the soy supplemental feeding program and the development of *Bt* cowpea in Ghana, discussed in Chapters 2 and 3, respectively.

One of the book's central concerns is how the new Green Revolution for Africa expands the reach of this governance of humanitarianism by way of philanthrocapitalist development. As discussed in Chapter 2, democratic accountability and transparency is not required of philanthropists funding development interventions. Given evidence of donor-driven tendencies and inattention to metrics beyond business performance as well as the scope of philanthrocapitalist development projects, this lack of accountability and transparency is problematic. Such patterns raise questions regarding the power dynamics between donors and recipients, as well as how these relationships *empower* certain actors (philanthropists get to feel good in "doing good" and their monopoly on wealth is better legitimated) while *undermining* the agency of others (e.g., the logic of "beggars can't be choosers" and "we know what's best for them").

In critiquing what she terms the "long shadow of Band Aid humanitarianism," Tanja Müller (2013: 470) argues that the representation of famine in Band Aid campaigns "was instrumental in establishing a hegemonic culture of humanitarianism in which moral responsibility towards impoverished parts of an imagined 'Africa' is based on pity rather than a demand for justice." This kind of mobilization of empathy by way of celebrity appeals "elicits 'the fantasy of a global moral community' at a time when global inequalities have reached unprecedented levels" (Müller 2013: 471). However different than celebrity-driven strategies to address famine, calls for a "new Green Revolution" frequently invoke "starving Africans" to emphasize its necessity (e.g., Rodin, as quoted by the Bill & Melinda Gates Foundation 2006; World Bank 2007). Such representations do violence in their objectification and simplification, and undermine the agency and capacity of African people who *can* and *do* find solutions to their own problems.

This question of agency is what distinguishes food sovereignty from food security. Food security is commonly understood as existing "when all people, at all times, have physical and economic access to sufficient safe and nutritious food that meets their dietary needs and food preferences for an active and healthy life" (FAO 2006). Although it is cognizant of preference and the need for "safe and nutritious food" alongside continuous physical and economic access to it (reflecting Amartya

Sen's entitlement approach to the study of poverty), it misses the emphasis on the autonomy and control of communities in deciding their agricultural futures as well as "the right of peoples to … culturally appropriate food" emphasized within a food sovereignty approach (e.g., Declaration of Nyéléni 2007). That is to say that food security achieved through "supplemental nutritional feeding" and food aid reflects a lack of agency among recipients that are just that, there to receive what is given to them.[27] This is akin to what Escobar (1995) refers to as "encountering development," rather than guiding it. By contrast, food sovereignty is about peoples' ability to determine how food should be produced, distributed, and consumed.

Another concern of this book is how the character of state agency and authority has changed under neoliberalism and is further shaped by the rise of philanthrocapitalist development. In her seminal work, *The Retreat of the State: The Diffusion of Power in the World Economy* (1996), Strange asserts that it is imperative to look at the power exercised by authorities other than states. She argues that there has been a growing diffusion of authority since "now it is markets which, on many crucial issues, are the masters over the governments of states" (Strange 1996: 4). The paradox is that whereas there has been an overall decline in the authority of states, there has been an increase in government intervention (ibid.). In other words, there has been a retreat not in the *quantity* of authority exercised by states but in the *quality* of that authority (Strange 1996: xii). That qualitative change is captured by Foucault's articulation of the central dictum of the neoliberal state: "One must govern for the market, rather than because of the market" (Foucault 2008: 121).

Governance for the market rather than because of it is reflected in the work of the "enabling state" that both reduces risks for the private sector in order to attract investment (often by taking on their own risks and investments (McGoey 2014)) and delegates some of its responsibilities. This shift is further observed in the emergence of public–private partnerships (PPPs) that distribute responsibility among actors and have expanded over the past decade in their role in the provision of public goods. Ferguson (2006: 38) highlights how donor policies that shift funding away from African bureaucracies and towards NGOs as more "'grassroots' channels of implementation" have led to many functions of the state being "effectively 'outsourced' to NGOs" with the consequence of the deterioration in state capacity. Moreover, this diffusion of responsibility makes it difficult for actors to be held accountable in cases of injustice experienced by PPP- and NGO-led projects—thereby undermining the agency of affected parties to pursue a just remedy.

Chapter 6 conceptualizes agency differently: by considering the "neocolonial anxieties" of diverse actors in Ghana preoccupied with the role of external influence on food and agricultural policy in Ghana. My work suggests that we need to avoid overly deterministic views of the subjugation of the postcolonial state to outside interests and to consider "vectors of neocolonialism" that may spread through the economy, ecology, and legal environment. The meaning of "vector" refers not only to the idea of an organism transmitting a pathogen but also to magnitude and direction, and allows space for contingency in its effects.

In doing so I consider a notion of distributed agency that takes into account the agency of nonhuman forces (Bennett 2010)—from transgenes to fall armyworms to climate change—that also shape Ghana's agrarian future.

Methodological Approach

Contesting Africa's New Green Revolution critically interrogates the claim that the new Green Revolution for Africa interventions are pro-poor and reveals the ways in which such agricultural interventions may, by contrast, exacerbate the inequalities that they were intended to address.

In doing so, it contributes to debates central to a school of critical development studies exemplified in *The Development Dictionary: A Guide to Knowledge as Power* (Sachs 2010a) that works to denaturalize notions of progress and development, and conceives of modernity as a culturally particular construction that reflects Western values and assumptions. It does so both to highlight science and technology as historical and cultural productions (e.g., Sunder Rajan 2006; Foucault 1990) and to challenge triumphalist and universalist Green Revolution discourses.

My methodological approach is situated within interpretive[28] social science that views knowledge, including scientific knowledge, as historically situated in power relations. It is guided by the insight that discourse itself has political and material effects, and interrogates what identities and agency as well as forms of knowledge and property are mobilized within the discourses that circulate the new Green Revolution for Africa agenda. The project uses a blend of methods—discourse analysis, participant observation, and interviews—in order to analyze debates about food security, agricultural modernization, and the cultivation of GM crops in Ghana. I complement discourse analysis with rich locally specific observations gained by both participant observation of farming and agricultural research communities and extensive interviews with a wide range of knowledge holders in Ghana between December 2011 and August 2018. This combination of methods is well suited for the analysis of power struggles that influence the trajectory of this agricultural transition.

In studying the discourses that animate, enable, justify, or resist agricultural change in Ghana, I employ a Foucauldian discourse analysis that is influenced by the post-development literature (Escobar 1995, 2008; Sachs 2010a; Rahnema and Bawtree 2008). A Foucauldian approach to discourse analysis recognizes that discourse itself exhibits agency: discourse makes certain things easier to imagine and other things harder to question.[29] Discourse analysis is concerned with the study of sets of articulated practices that unite around a common set of meanings, values, and perspectives, "the process through which social reality comes into being ... the articulation of knowledge and power, of the visible and the expressible" (Escobar 1995: 39). Discourse analysis is useful for studying assumptions regarding cause and effect, what actors are considered legitimate stakeholders, what kinds of knowledge and values are privileged, and how key contested terms like "sustainability" are framed.

Discourse analysis of the GMO debate—central to understanding the politics of the new Green Revolution for Africa—allows me to situate my research in Ghana within global dynamics that identify certain populations and places as deficient[30] and in need of intervention. Discourse analysis also reveals how both proponents and opponents of GM crops alike place *life at the center* of the debate—albeit vastly different conceptions of life. Discourses prescribe certain practices that work to shape social reality. For example, the discourse of progress places great value on science-based solutions and articulates a vision of the future in terms of consistent improvement. When the adoption of new agricultural technologies is framed as part of "progress" and "modernity," it produces corollary subjects that are "modern." Likewise, the rejection of such technologies can render those that are averse to these technologies as "backwards" (Ignatova 2015).

This approach allows me to study how identities such as the entrepreneur are shaped through language and produced and reproduced through agricultural practices. Because the expansion of plant biotechnology is conditional on information flows, information regarding GM crops must be disseminated in order to both ameliorate anxieties of the public regarding this new technology and train farmers and researchers on how to use it given biosafety protocols. The way in which this information is distributed, framed, and positioned in the Ghanaian context is critical to my analysis. I pay attention to how issues are linked, how means and ends are defined, and what identities are being promoted (e.g., agrarian entrepreneur vs. self-sufficient farmer).[31] For example, the framing of food production in African countries as a system in crisis has made technologies like GM seeds that promise higher yields increasingly attractive and more likely to be adopted. Yet activists' counter-frames of GM seeds as threatening motivate calls for outright bans of this new technology.

My fieldwork consists of on-the-ground field research in northern Ghana conducted during four trips: December 2011–January 2012, July 2012–March 2013, May 2015, and July–August 2018.[32] While in the field, I participated in farming and postharvest activities in rural communities in the Upper East and Northern Regions of Ghana. I conducted interviews with a range of actors across northern Ghana and in Accra—farmers, farmer organizations, traditional leaders, policymakers, actors working for aid agencies and NGOs, activists, scientists, academics, bureaucrats, and actors within agribusiness—about how they perceive agricultural challenges in Ghana and what roles the state, private actors, new technologies, and local communities can play in addressing them. Through repeated travel to Ghana, I was able to gain greater access to a wide range of stakeholders (including elites that would be otherwise difficult to reach) and collect richer accounts of the politics of agricultural development in Ghana. I interviewed both advocates and opponents of biotechnology, discussed their informational strategies, and obtained samples of the materials they circulate to advance their positions. I also interviewed the main actors involved in the recent court case brought by Food Sovereignty Ghana against the Ghanaian government to halt the commercialization of GM crops as well as those food sovereignty organizations that existed prior to the passage of the Ghana Biosafety Act.

By combining field observations with discourse analysis, I situate my study simultaneously at the local and global levels. That is to say that my findings are derived both from particular observations on the ground in Ghana and by situating what is happening in Ghana within a broader political-economic and discursive context. Such an approach is inspired by Anna Tsing's (2005) "ethnography of global connection." It should be stressed that I am studying a period of agricultural transition as actors within Ghana begin, resist, and/or consider the adoption of GM crops and new modalities of farming. The purpose of this book is not to generate conclusive findings, but to *provide a lens through which to understand currently unfolding ideas and practices.* The analysis of such a transition contributes to the study of the micro-practices of development change (Escobar 1995). A challenge to studying a transition is that information is inherently incomplete and the motivations of development may be obscured. Yet, the posturing, the framing, and the blocking out of alternative truths that are part of this information production in themselves produce potent political effects that shape peoples' livelihoods and local ecologies (Tsing 2005).

Structure of the Book

Chapter 1, "Green Revolution Discourse, Structural Adjustment, and the 'Enabling Environment' for Agribusiness," begins with the question, how new is this new Green Revolution for Africa? It underscores that the Green Revolution did not "miss" Africa; rather, the development interventions were either not sustained or were inappropriate (Akram-Lodhi 2013; Patel 2013; Wiemers 2015; Goldman and Smith 1995). The chapter also considers the political benefits of framing these efforts to "transform" African agricultural systems as "new" and "uniquely" promising, arguing that this framing obscures the role that some of the nGR's biggest proponents played in creating the conditions for its supposed necessity. The narrative that struggling agrarian livelihoods in Africa primarily reflect problems of poor production ignores the history of bad advice that has been disseminated to African governments as conditions for the receipt of World Bank or International Monetary Fund loans: namely, pressure on African states to slash spending in agricultural research and development as well as extension, to end fertilizer subsidies, and to strive for a balanced budget at all costs.[33] This form of amnesia is akin to what Mitchell (2002) describes in the *Rule of Experts* whereby USAID and other development planners leave their own policy impacts outside of their analysis and depoliticize problems of food insecurity, poverty, and inequality.

Chapter 1 compares the first and the new Green Revolution, underscoring the geopolitical as well as biopolitical dimensions of these agricultural modernization programs. It also highlights changes in the role of the state in agricultural development: shifting from manager to "enabler" during the 1980s African economic crisis when African countries sought the financial assistance of the World Bank and International Monetary Fund (Holmén 2005: 93), with the state now an "enabling" partner in neoliberal PPPs. It undertakes a discourse analysis of the term "enabling environment" that has been constantly present in interviews, meetings, World Bank documents, and governmental statements regarding how to increase investment in developing countries. It also examines how the term distributes agency and diffuses responsibility, as well as the shift in agency and authority of the state that is corollary to the work of the creation of this enabling environment. I argue that such a shift in understanding of the role of the state disempowers it, even as it expects *more* from the state in its work to lure investment—governing *for* the market, rather than *because of* it (Foucault 2008).

Chapter 2, "Philanthrocapitalism and the Politics of Public–Private Partnerships," explores what is at stake in agricultural development objectives achieved by way of philanthrocapitalist PPPs, a defining feature of the new Green Revolution in Africa (Thompson 2014; Moseley 2017). When PPPs are philanthrocapitalist, they share the assumption that gains for both capital and the poor alike can be achieved through development by way of "impact investing" that supports entrepreneurship and requires measurable outcomes. Yet PPPs are neither neutral arrangements nor pro-poor. They reflect a form of development that is conditioned on the "win-win": that the gains of development are explicitly shared between the public and private sectors. The chapter's central question is *who benefits from philanthrocapitalist development in Africa?* This analysis suggests that philanthrocapitalism primarily serves capital, with philanthropy being a byproduct of the pursuit of powerful actors' interests. Philanthrocapitalism can aid in the pursuit of aims by governments but, given its donor-driven nature, may serve donors far better than recipients. I argue that the increasing prominence of philanthrocapitalist-led development in Africa raises important concerns about accountability, power, and democracy. Furthermore, its expanding role can also be understood as part of the growing governance of humanitarianism (Barnett 2011).

My analysis also examines how philanthrocapitalist PPPs are shaping agricultural diets and futures in Ghana. I define public–private partnerships as the public sector provision of an enabling policy environment for the involvement of private companies, NGOs, foundations, and foreign aid agencies in the financing, design, construction, operation, and ownership of a public sector service or utility. The idea of PPPs fits within neoliberal discourse that promotes smaller government and a growing role for the private sector, but it also emerges in the wake of realizations that privatization has been "oversold" (e.g., Ostry, Loungani, and Furceri 2016). As an example, I briefly highlight both how philanthrocapitalist PPPs "cultivate taste" for soy in Ghana by examining the work of distributing certified soybean seed to farmers, supporting soy supplemental nutritional feeding in schools, and promoting soybean as feed for the poultry sector and analyze the political-economic consequences of doing so. This case study provokes the question: to what ends is a commercial soy market being promoted?

Whereas Chapter 1 shows that the idea that the Green Revolution missed Africa is a myth that obscures how many Green Revolution practices have been introduced in Africa but not sustained, Chapter 3, "Biocapital, 'Pro-Poor' Biotechnology, and Legislative Changes in the Seed Sector," argues that there is, in fact, something new about this new Green Revolution. This powerful assemblage, whose objective is to transform African agricultural systems, has contributed to the proliferation of new forms of capital—biocapital and philanthrocapital—that integrate biotechnology with philanthropy to create market value. Chapter 3's discussion of biocapital contributes to literature on the bioeconomy and builds off of Escobar's (2008) notion of "genecentrism." It explores how the expansion of capital is engendered by philanthropic giving that normalizes seed as commodity and legislative reform that renders seed patentable material. Through case study and discourse analysis of efforts to develop a GM legume, *Bt* cowpea in northern Ghana, I look at how one emblematic philanthrocapitalist move—Monsanto's "donation" of the *Bt* transgene for the development of pro-poor biotechnology—can be seen as a mechanism to normalize a Western patent system and techno-scientific agribusiness model that depends upon expertise from elsewhere.

Chapter 3 also highlights legislative changes in Ghana that indicate shifts in how plants and seeds are understood. For instance, the notion that seed is patentable material, previously part of an excluded category in the Ghana Patents Act of 2003, is gaining some acceptance within policy circles. This is indicated by current discussions around the Plant Breeders' Bill (renamed the Plant Variety Protection Bill) recently passed by Ghanaian parliament, in line with global pressures to standardize plant breeders' rights and intellectual property protections. Upon presidential assent, it would strengthen the rights of foreign and select domestic plant breeders. A growing acceptance of the patentability of plants indicates that the property rights regime in Ghana is shifting away from complex customary notions of property toward Western neoliberal ideas about ownership. Philanthrocapitalism facilitates this change. The understanding of the commodity form of seed is further reinforced through specific legal regimes that recognize intellectual property and grant greater authority to those actors that possess biosafety and biotechnology expertise. Moreover, the agricultural model described in Chapter 3 illustrates the elevated importance of the laboratory and the legal arena—rather than the farmer's field—to agricultural development.

GM seed is a uniquely polarizing technology. Chapter 4, "Technological Savior or Terminator Gene? Biotechnology, Food Security, and the Political Economy of Hype," showcases these contrasts between proponents that view the wholesale rejection of GMOs as needlessly shutting the door on a technology that could save lives and opponents that view this rejection as necessary in order to affirm food sovereignty and protect farmer livelihoods. GM seeds can represent either liberation—African scientists being able to produce pro-poor biotechnology responsive to communities' needs—or domination—ensuring the corporate control of the food supply through the "Trojan horse" of Monsanto seed product dependency. I contend that the polarized nature of debates over GM seed that frames African agriculture as a system in crisis makes way for exclusionary forms of agricultural development dominated by both experts and entrepreneurs. Opponents view GM seeds as a technology that poses unique risks, thereby requiring moratoriums, extensive study, and regulation of GM seeds. Proponents, in contrast, view biotechnology outreach, training, philanthropy, and stewardship over GM seeds as necessary in order to realize the benefits of this essential technology. This emergency framing makes piecemeal approaches, compromises, and deliberation appear irresponsible in the face of such urgency.

The fourth chapter considers how both proponents and opponents invoke both hype and science in this debate over GM crops. Drawing upon Tsing's (2005: 57) notion of "economy of appearances," I show how pioneering industries and campaigns are more inclined to use hype, or in Tsing's terms "spectacle," during critical times when support is most needed. Biotech proponents use hype at the research and development stage of new products or when products are criticized. Activists use hype during critical political economic shifts, like impending legislative changes on GMOs, or prior to the introduction of new transgenic products into commercial markets. Hype can be an important tool in garnering the support necessary to successfully ban GMOs or, by contrast, in generating the capital for new expansions in biotechnology research. However, for those opponents who view GM seeds as a symbol of deeply rooted problems of inequality, power, and injustice, the emergency framing—through its consequent shifts in authority and its narrowing of focus—may hinder efforts to address these social problems. It may do violence to small steps and long-term strategies of individual farmers and communities to pursue homegrown, endogenous solutions.

Between 2009 and 2019, both endogenous and exogenous development planners in Ghana hyped up the idea of "farming as a business" on the basis that such a strategy could benefit Ghanaian agribusiness and alleviate poverty simultaneously. Chapter 5, "Experts, Entrepreneurs, and the 'Last Mile User,'" examines what livelihood practices would change if such idealized notions of agricultural transformation were implemented. Such a consideration of the meaning and practices of "from farming as a way of life to farming as a business" is based on participant observation of farming, international development, and agricultural expert communities in northern Ghana (Tamale and surrounding villages, Nyankpala, Bolgatanga, and Bongo and surrounding villages since December 2011). The chapter highlights the central figure to enact this transition from farming as a way of life to a business: the entrepreneur that is linked with experts, opportunities to professionalize, and markets.

The chapter explores this discourse of entrepreneurship as well as the nexus of relations between experts, managers, entrepreneurs, and the "last mile user"—the farmer that adopts modern agricultural technologies and integrates into outgrower contracts. I argue that the new Green Revolution for Africa interventions normalize a particular modality of farming that may introduce new vulnerabilities (e.g., expensive input packages whose formulaic nature does not respond well to deviation, reliance on irrigation systems that may not function, and debt) as it mitigates others (e.g., securing contracts, attracting investment, and making technologies more accessible).

Chapter 5 details how a commercial seed sector is being developed[34] in northern Ghana by way of development expertise, entrepreneurship, and philanthrocapitalist development that standardize and normalize agricultural production. It reveals a contradiction of the new Green Revolution for Africa discourse: this agenda is presented as pro-poor, but in reality it is "pro-private sector" driven, working to expand a consumer base of "last mile users." In particular, it reflects upon USAID's Agriculture Technology Transfer (ATT) Project's last mile user delivery model that considers farmers in these market frontiers as "users," rather than generators of technology and holders of seed. Further upstream is the role of development experts and universities that generate the products that feed into these projects, such as the University of Illinois Soybean Innovation Lab's "soybean success kit," and training such as the MS degree in Seed Science and Technology at the West African Centre for Crop Improvement that "leaves the entrepreneurial class to do the work of seed development ... managing companies, research stations, seed production."[35] The chapter argues that the development of an entrepreneurial-managerial class works hand in hand with privatized regulation and plant breeding to attract global investment.

The interlude, "On 'Mixing,'" serves as a challenge to the totalizing narratives, on the one hand, of proponents of the agricultural transformation agenda that claim it "will help tens of millions of people who are living on the brink of starvation in sub-Saharan Africa" (Bill & Melinda Gates Foundation 2006) and, on the other, of food sovereignty activists that sometimes romanticize peasant farming and advocate for farming systems free from industrial influences. Through a series of stories, this brief section illustrates the practice of pragmatic "mixing" as farmers pick and choose what kinds of "modern" or "traditional" agricultural practices to adopt, playing multiple roles that secure not only income but also food for their families." Such practices of "mixing" reveal how even some of the farmer entrepreneurs that are selected to lead the commercialization of agriculture hold tight to tradition—complicating both the agricultural and identity shifts perceived to be crucial components in efforts to transform farming from a way of life to a business in Ghana.

Chapter 6, "Neocolonial Anxieties," uses fieldwork to highlight the shared neocolonial anxieties regarding outside influence on food and agricultural policy in Ghana. Drawing upon Kwame Nkrumah's theorizing on neocoloniality, I explore what I term different "vectors of neocolonialism": ecological, economic, and legal. To illustrate an ecological vector of neocolonialism, I describe the forces set in motion by the "invasion" of the fall armyworm in Africa—including a new expert rationale for GM crops—and the environmental conditions that foster its spread. The "slow violence" (Nixon 2011) of weakening institutions of plant research and breeding—an effect of structural adjustment—is theorized as an "economic vector of neocolonialism" that incentivizes PPPs, that, rather than producing more equitable and pro-poor relations, enable outsized donor influence on agricultural priorities in Ghana. Legal reform in the domains of plant breeding, intellectual property rights, and biosafety suggests that legislative change's greatest beneficiaries may be global agribusiness; this vector of neocolonialism has the potential to create new economic relationships to global agribusiness in the seed, agro-chemical, and food markets that may come at the expense of local agribusiness.

Central to Chapter 6's examination of contemporary neocolonial politics is a critical consideration of what food sovereignty—which I define as the ability for nations and communities to determine how and what food should be produced, distributed, and consumed—looks like in the Ghanaian context. Drawing upon Agarwal (2014), I utilize my fieldwork to illuminate the tensions between democratic decision-making and popular notions of food sovereignty. I conclude with a series of questions concerning food sovereignty and the decolonial project for the consideration of scholars and activists alike.

Contesting Africa's New Green Revolution contributes to the analysis of the "politics of poverty" of the new Green Revolution for Africa agenda (Escobar 1995: 23; Spann 2017; Moseley, Schnurr, and Bezner Kerr 2015), and the conclusion both seams together its contributions to this area of study and highlights future research trajectories. It further raises political-ecological concerns about the implementation of new Green Revolution programs in Africa in terms of the high levels of fossil fuel–based inputs and irrigation used—to say nothing of the fossil fuels associated with transport in global value chains. Strategies like these that contribute to the climate crisis can no longer be acceptable as the effects wreak havoc on agrarian livelihoods on the climate frontlines,[36] including some of the communities I visited targeted for new Green Revolution interventions. The conclusion addresses the question of alternatives through an exploration of forms of community agroecological governance and seed stewardship, local philanthropy and assistance, and knowledge production that subverts the logic of extroversion that, together, can enable genuine food security and food sovereignty.

In sum, the book speaks to literature in critical development studies and the politics of poverty by its emphasis on how international development institutions and funders of Green Revolution programs distance themselves from their own roles in disseminating policy advice that undermined farmers' livelihoods. Following analysis of the Green Revolution and the new Green Revolution for Africa—and whether it is really *new* or *for* Africa—I look at how philanthrocapitalists shape the landscape and logic of development in Africa today. I examine how the win-win logic of philanthrocapitalism that frames agricultural biotechnology as pro-poor helps to facilitate a controversial technology into African frontier spaces and reflects a governance of humanitarianism that privileges donor interests over recipient priorities. Part of what mobilizes philanthrocapitalist donor interest is through what I term a "political economy of hype": the deployment of simplistic, exaggerated claims to stimulate activity such as consumption, investment, and philanthropy or to mobilize activism. The book highlights how the promise and potential of the poverty-alleviating properties of new technologies are sufficient for capital's expansion but may consistently underdeliver in their promises for the poor. I consider hype in the context of the agricultural transformation agenda in northern Ghana: how access to technologies and training combined with changes in farmers' "mentality" (ACDI/VOCA n.d.) are supposed to solve the problems of rural development in northern Ghana. The book looks at how the identity construction of the serious farmer blames smallholder farmers for their own poverty and conceals how these new Green Revolution interventions, like many before, serve best elite and global agribusiness interests. Finally, I develop a framework of analysis, vectors of neocolonialism, to understand the dynamics that serve to reproduce the conditions of this politics of poverty.

Chapter 1

GREEN REVOLUTION DISCOURSE, STRUCTURAL ADJUSTMENT, AND THE "ENABLING ENVIRONMENT" FOR AGRIBUSINESS

I. Introduction

Florence Wambugu, a Kenyan plant pathologist and virologist, the CEO of Africa Harvest Biotech Foundation International, and a vocal proponent of genetically modified (GM) crops, makes a claim common to proponents of GM crops and a "new Green Revolution for Africa": "We may have missed the green revolution, which helped Asia and Latin America achieve self-sufficiency in food production, but we cannot afford to be excluded or to miss another major global technological revolution" (Wambugu 2001). Suggested here is that it is the absence of these technological revolutions that accounts for African countries' supposed inability to be self-sufficient food producers. As such, a new Green Revolution for Africa appears to be self-evidently positive and necessary and a means for better inclusion in the technological changes happening globally.

The "Green Revolution" is a referent to the high crop yields that countries in Asia and Latin America experienced in the mid-20th century through the introduction of scientific seeds, agro-chemicals, irrigation, and linkage to markets, and also suggests a benign, if not positive, effect on the environment. This "new Green Revolution" (hereafter nGR) is a political project advanced by a global assemblage of actors that includes governments, international and bilateral aid agencies, foundations, international and national agricultural research institutions, and global agribusiness corporations. However, efforts to modernize agriculture in Africa are not new, and the idea that reforms in small-scale agriculture can lead to a "transformation" in African economies has been repeatedly invoked in African agricultural policy since the 1960s (Wiemers 2015: 104; Amanor 2009; Holmén 2005). *What is new is who leads*: whereas in the 1960s the state led efforts to modernize agriculture, now public–private partnerships take the lead (Amanor 2009: 247; Moseley 2017; Brooks 2005: 362). I discuss the composition and power of these public–private partnerships in the following chapter.

I consider two effects of the framing of this African Green Revolution as "new" and necessary. First, I look at how this narrative of an excluded Africa views the limited agricultural research and development in Africa as a product of the absence of delivery of technical products from the West, rather than policy failure. Second, the idea that the Green Revolution "missed" Africa obscures how national agricultural development programs were undermined by some of these same actors promoting the new Green Revolution for Africa agenda. This chapter not only examines the political utility of framing these interventions as novel and necessary but also evaluates what, if anything, is new about the new Green Revolution. This query is further expanded in Chapters 2 and 3. Central to this analysis is the shifting roles of the public and private sectors in agricultural research and development in Ghana.

The chapter is organized as follows. First, I examine the key characteristics of the first Green Revolution and highlight its dominant framing of agricultural development as a problem of production. I also critique the notion that the first Green Revolution missed Africa. Second, I discuss the changing role of the state in agricultural research and development to that of an "enabler," noting shifts both in funding and in the roles of the public and private sectors. Then I connect the legacy of structural adjustment to the logic of the new Green Revolution for Africa by questioning the origins of the deficiencies the nGR agenda is intended to address. I identify a number of key continuities between the two "revolutions" that have been obscured by their depoliticized presentation as a technical endeavor. In the final section I analyze the language of "enabling environment" that is frequently invoked by development planners and actors advancing a new Green Revolution and show that this discourse reveals continuity with the rationale of structural adjustment. I conclude with a critique of the new Green Revolution agenda that reveals it to be an ahistoric productivist lens that views problems of African agriculture as fundamentally issues of production and that ignores the past roles that its proponents played in promoting policies that have undermined African agricultural development.

II. *The First "Green Revolution": An Overview*

The first Green Revolution is characterized by a technological breakthrough in the development of hybrid varieties of cereal grains; increased mechanization, irrigation, and fertilizer application; the growth of international and national agricultural research institutes; and the geopolitical context of the Cold War (Oasa 1987: 40; Griffin 1974; Perkins 1997; Evenson and Gollin 2003). The term "Green Revolution" was coined by William Gaud, former director of USAID, to contrast it with a "Red Revolution" of the spread of communism or a "White Revolution'" of land redistribution in Iran (Patel, Holt-Giménez, and Shattuck 2009).[1] Supported by the Rockefeller and Ford Foundations and the US government as a means to stop the spread of communism, Green Revolution programs initiated in Mexico in 1941 and India in 1956 were designed to promote high-yielding agricultural practices. Faced with a plant disease that led to a significant decline in yields, Mexico was targeted as the first site for the Rockefeller Foundation's agricultural assistance under the leadership of biologist Norman Borlaug. This choice was strategic given that the Roosevelt administration "wanted neither a socialist nor a fascist state on its southern border" (Perkins 1997: 9). The first Green Revolution was a geopolitical project, not merely a technical endeavor.

The Green Revolution promotes "plant improvement" and crop productivity as central to development and has been supported by international agricultural research centers such as the International Rice Research Institute (IRRI), the International Maize and Wheat Improvement Center, and the International Institute for Tropical Agriculture (IITA); bilateral aid agencies; the Rockefeller, Ford, and Kellogg Foundations; agribusiness corporations; and national agricultural research institutions. The Green Revolution programs introduced high-yield dwarf varieties of wheat and rice and hybrid maize as well as pesticides, fertilizers, and improved irrigation technologies. This capital-intensive agriculture was at the time attributed with averting famine in India and helping Mexico become an export-oriented agricultural economy; later these agricultural practices were linked to ecological degradation and social disparities in the agricultural sector (Weissman 1990; Shiva 1991; Patel 2013; Freebairn 1995; Sobha 2007). High-yield varieties of rice, wheat, and maize were developed in Asia and Latin America suited to their respective local agroecological context. By contrast, Asian varieties of grains were either brought to African countries without similar adaptations or prioritized traits that required formulaic and time-sensitive applications of "seed-water-fertilizer" in order for yields to increase (Berry 1993: 199). Both approaches generated disappointing results in the African context (e.g., Evenson and Gollin 2003).

The first Green Revolution aimed to achieve agricultural transformation through a change in process as it enabled a new method of producing particular crop commodities (Griffin 1974: 48). In order to realize the benefits of these high-yielding varieties, a new process of the use of supportive agricultural technologies—fertilizers, pesticides, and irrigation—changed agricultural systems and the landscape upon which they were grown. In addition to creating a commercial seed sector, this shift toward input-intensive agriculture led to growth in agro-chemical, farm machinery, irrigation supply, and other supporting sectors. The Green Revolution worked to standardize agricultural practices and generate surplus through the adoption of more limited number of cereal varietals and a more formulaic application of inputs. This standardization in process rendered legible agricultural systems for global markets.[2]

One of the overriding concerns of the Green Revolution was the overall increase in agricultural productivity measured in aggregate terms, and this was achieved in certain staple food grains. In India, whose agricultural programs during this period were frequently given as a Green Revolution success story, yields for rice increased from 902 kg/ha in 1953–4 to 2,240 kg/ha in 2010–11, whereas those for wheat increased from 750 kg/ha in 1953–4 to 2,938 kg/ha in 2010–11 (Basu and Sholten 2012: 111).[3] It should be noted, however, that there is much debate over the reliability of statistics that demonstrate a "revolution" in agricultural productivity in India, although there is greater consensus that there were agricultural yield gains between the mid-1960s and early 1970s (Jervens 2014; Rudra 1982). Such increases in yield in these cereal grains were generated through the "productive power" of industrial inputs that substituted (and undervalued) the "reproductive power" of nature (Yapa 1996a: 82; Shiva 1991; Sobha 2007). The components of farming reliant on the reproductive power of nature—seed saving, rainfall, composting, intercropping, the use of manure and animal labor—were substituted by the purchase of scientific seeds, the establishment of irrigation systems, and the application of fertilizers and herbicides. Furthermore, "women and peasants who were traditional agricultural experts" were displaced by techno-scientific expertise that generated and disseminated Green Revolution technologies (Sobha 2007: 107; Richards 1985).

This introduction of Green Revolution technologies marks a shift from a labor-intensive agriculture dependent upon the reproductive power of nature and community labor to a capital-intensive system dependent upon industrial inputs and mechanization (Yapa 1996a; Shiva 1991). In this sense, Green Revolution technologies were landowner- and elite-biased, offering differential results to peasant farmers that both did not have access to capital and were considered too high-risk to have access to credit. Griffin (1974: 30) argues that "unequal access to land and capital frequently is accentuated by unequal access to water and technical knowledge." In other words, landed wealth frequently goes hand in hand with political influence and, with that, privileged access to scarce means of production (i.e., tractors, subsidized fertilizer, the most fertile land) (Griffin 1974: 18). These existing patterns of inequality were deepened by technological change and government policies that privileged improvements in wealthier farming regions like rich river valleys that generated aggregate productivity gains, rather than improving the conditions at the household level through attention to peasant farmers reliant on rain-fed agriculture. Large-scale commercial farmers were given subsidies to mechanize production, whereas peasant farmers, especially women, struggled to gain access to small amounts of credit to improve their farming (Griffin 1974).

The First Green Revolution in Africa

Between the 1960s and the 1980s, the Ghanaian government implemented policies of self-sufficiency in food production through subsidizing inputs such as fertilizers, providing support for marketing, a system of guaranteed prices for crops such as rice, and access to credit (Mittal 2009: 24–5; Berry 1993: 56–7). The Ghanaian government also experimented with state farms and large-scale irrigation projects as well as expanded farmers' cooperatives (Berry 1993: 57). The Nkrumah administration recognized the need to diversify crops, expropriated land for state-run farms, and "drew resourcefully on ideas and aid from both the East and the West" to increase agricultural production, however unsuccessfully (Lambert 2019: 44). In 1972, the military dictatorship of Colonel I. K. Acheampong instituted an ambitious national self-reliance plan entitled Operation Feed Yourself. The emphasis on large enterprises and export production, at the neglect of smallholder farmers and domestic demands, failed to produce expected results (Girdner et al. 1980). Yet, as Wiemers (2015) articulates in her study of Kpasenkpe, a town in the Northern Region of Ghana, small-scale farmers were also beneficiaries of state-subsidized fertilizers and new varieties of maize, a historical period (1966–early

1980s) her informants called "a time of Agric" when support from agricultural extension was notably present.

Like Ghana, the 1970s signaled for most African countries a shift toward programs of national food self-sufficiency in order to address a growing population as well as the implementation of Green Revolution measures into their agricultural policies (Holmén 2005: 88; Wiemers 2015; Goldman and Smith 1995). In 1980, the Lagos Plan of Action, "borne out of an overwhelming necessity to establish an African social and economic order primarily based on utilizing to the full the region's resources on building a self-reliant economy," was initiated that included commitments to "achieve self-sufficiency in food production and supply" (UNESOC 1991: 1; Organization of African Unity 1980: 4). African governments established state farms, large-scale irrigation programs, marketing boards, input subsidy programs, and minimum price guarantees. Much of the focus of these agricultural programs was placed on large estate agriculture that targeted large "progressive farmers" who received subsidized fertilizer and other inputs, low-interest loans, and tractors (Amanor 2011: 50). Land reform—only mentioned once briefly in the Lagos Plan of Action—was not a focal point of these initiatives, allowing colonial-era land access problems to persist.[4]

A number of countries were self-sufficient in food crop production during this time and production of maize and rice increased but often as a result of the expansion of cultivated land rather than yield increases (Patel 2013: 33; Evenson and Gollin 2003: 760; Goldman and Smith 1995). Goldman and Smith's (1995: 254–6) study of northern Nigeria indicates that Green Revolution practices were evident beginning in the late 1960s with the widespread adoption of maize developed by the IITA and the national Institute for Agricultural Research, the increased utilization of fertilizer, as well as practices of "extensification" that cultivated more land through oxen plowing and increased use of hired labor. They also note that these shifts in practices were stimulated by substantial state support in the form of the expansion of road networks, the development of farmer supply centers, as well as fertilizer subsidies.

The introduction of Green Revolution technologies in the African continent can be characterized as inappropriate[5] in the case of hybrid seeds and inconsistently accessible in the case of fertilizer. I will take these two points in turn. Hybrid varieties of crops required specific application of inputs that were too inflexible for the realities of African farming:

> More generally, the technical requirements of Green Revolution seed-water-fertilizer packages have often conflicted with African farmers' need for flexibility in managing their time. Many high-yielding varieties of seed have longer maturation periods than traditional varieties. They perform best when grown in sole stands rather than intercropped fields, and their yields are more sensitive to the timing of labor inputs than those of traditional varieties. (Berry 1993: 199)

Berry argues that African farmers need greater flexibility both as a response to economic volatility (that increases the incentive to keep options open) and due to the declining ability of African farmers to recruit sufficient labor during key cultivation tasks. As such, when farmers tried hybrid varieties, they were likely not able to meet the prescriptive growing requirements of these new crop varieties, which resulted in unsatisfactory outcomes that discouraged its adoption. Fertilizer use, on the other hand, reflected problems of access as many farmers on the African continent could not afford imported fertilizer without government support (Martey et al. 2014). Farmers took advantage of subsidized fertilizer when present—often during election cycles when politicians sought to court public opinion—and cut its use when they could no longer afford it.

III. Structural Adjustment and the Impact of Austerity on Agriculture

In the first Green Revolution, the state played a more central role in plant breeding and crop improvement research through large amounts of funding for national agricultural research institutions. During this time, Ghana's national development model rhetorically embraced self-sufficiency; however, it never fully abandoned its emphasis on export crops. As such, global declining terms of agricultural trade put a significant part of the economy at risk. Ghanaian agricultural development policies reflected commitments to a productivist paradigm, rather than to the address of colonial-era problems of unequal access to land and resources. These vulnerabilities encouraged Ghana to seek financial assistance. The role of the state in agricultural development changed from being a manager to an "enabler" during the 1980s African economic crisis when African countries sought the financial assistance of the World Bank and International Monetary Fund (IMF) (Holmén 2005: 93). Support for African countries was given on the condition that programs such as input subsidies and guaranteed minimum pricing were reversed and state support for agriculture was cut significantly (World Bank 1981; Gibbon 1992). These institutions viewed Africa's economic crisis as the result of bad policy and excessive state intervention and spending (World Bank 1981; Berry 1993: 136; Mkandawire and Soludo 1999: 24).

Ghana began adopting structural adjustment in 1983, the same year that an acute drought greatly impacted the country, and agreed to its first structural adjustment loan in 1984 (Gibbon 1992: 64). The World Bank's reform efforts in Ghana in this period were focused almost exclusively on the transformation of the cocoa sector, which had underwhelming results due largely to low world prices for cocoa (Gibbon 1992: 60, 77; Berry 1993: 179). Beyond the attempt to expand cocoa production, the adjustment agenda in Ghana also focused on deregulation of its input supply and state marketing functions (Gibbon 1992: 71). In 1989 the World Bank commended Ghana's Economic Recovery Program (1983–6), Ghana's program of structural adjustment, for its "impressive" "achievements" (World Bank 1989b: i).

Under adjustment, "the only state involvement contemplated (cautiously) was the possible setting of minimum support prices for food crops, in order to create small buffer stocks for price stabilisation purposes" (Gibbon 1992: 53). The adjustment period was associated with sharp declines in agriculture's share of public expenditure as well as a dramatic downturn in input consumption due to cuts to government fertilizer and agro-chemical subsidies (Gibbon 1992: 76–7). Gibbon (1992: 58) anticipated that state cuts to fertilizer subsidies as part of structural adjustment will lead to producers ceasing to "purchase and apply that input," a prediction that has been borne out as development planners in Ghana today bemoan the low usage of fertilizer in Ghana.[6] Weissman's (1990) evaluation of structural adjustment programs (SAPs) looked to the Ghanaian and Kenyan experience to consider the impacts of structural adjustment, given the countries' reputations as "co-operative adjusters" that implemented economic reforms (Gibbon 1992: 50). Weissman served as a director of the United States House of Representatives' Subcommittee on Africa that examined whether SAPs promoted sustainable and equitable growth in sub-Saharan Africa. Weissman (1990: 1631) finds that the SAPs that Ghana undertook were often flawed given their reliance on short-term foreign consultants and thin on-the-ground research. Although international donors and recipient governments theoretically designed SAPs jointly, the role for national governments was weakest in sub-Saharan Africa (Weissman 1990: 1623; Patel 2013: 33).

Austerity measures in African countries contributed to a dramatic decrease in public-sector expenditures for plant breeding: expenditures dropped from 347 million in 1985 to 99 million in 2005 (Chambers et al. 2014: 28; Gibbon 1992: 59). Additionally, donor funding for agricultural research and development declined dramatically in the mid-1980s (Spielman, Zaidi, and Flaherty 2011). The Ghanaian government's agricultural spending in 2007 was 0.39 percent of total government expenditures, down from 12.23 percent in 1980 (Chambers et al. 2014: 27). During this time, agricultural research institutions were restructured to shift agricultural research to the private sector and the national agricultural research system was further integrated into the international agricultural research system, particularly the CGIAR network (Puplampu and Essegbey 2004: 275). This economic crisis led to a mass exodus of qualified staff from African countries, undermining the capacity of the national agricultural research institutes to conduct research, impacting their research agendas as well as the morale of agricultural research scientists (Puplampu and Essegbey 2004: 275, 277; Mkandawire and Soludo 1999: 135). Another manifestation of this underinvestment in agriculture in Ghana is agricultural extension: in 2012 the ratio of agricultural extension officers to farmers is 1:1,000, and in some cases the disparity is as great as 1:1,500.[7]

IV. The Need for a New Green Revolution for Africa in the Wake of Structural Adjustment

The 1980s economic crisis, drought, and the deterioration in African terms of trade for export crops contributed to the state's inability to continue food production policies focused on self-sufficiency (Holmén 2005: 89; UNESOC 1991), and SAPs encouraged this turn (Patel 2013; Wiemers 2015). The previous examination of agricultural policies in Africa in the 1960s and 1970s shows that Green Revolution programs didn't miss Africa as proponents of the new Green Revolution for Africa claim; rather, they were either not sustained (Djurfeldt et al. 2005; Patel 2013; Wiemers 2015; Goldman and Smith 1995; Akram-Lodhi 2013: 84) or ill-suited for the African context (Berry 1993; Richards 1985).

Drops in spending by both governments and donors on agricultural research and development continued to occur during the 1990s. A decade later, there was widespread recognition[8] that such spending cuts were detrimental to African development because of the central role of agriculture in African economies. In 2003, the African Union issued the Maputo Declaration on Agriculture and Food Security, which called on all African states to increase spending on agricultural research and development to a minimum of 10 percent of their national budget due to concerns that existing expenditure levels were stifling the development of African agribusiness (Fan, Omilola, and Lambert 2009). Ghanaian public sector expenditures have come close to meeting this goal in a single year (2013) with 9–10 percent of the total national budget spent on agriculture, but this goal has not been consistently met and averages about 6 percent a year (Curtis and Adama 2013: 7; NEPAD 2018).

Deficits of government investment in agricultural development create a demand for the private sector, aid agencies, and international agricultural research centers to fill in these gaps,[9] but as McGoey (2014) argues, attracting donor and philanthrocapitalist support necessitates government intervention to reduce risk. This dynamic is apparent with the decline in donor funding of African agricultural development concurrent with slashes to agricultural research and development spending by the public sector. The global food price crisis of 2007–8, combined with increased commitments to agricultural spending by national governments and foreign aid agencies, attracted greater private sector investment in agriculture (Spielman, Zaidi, and Flaherty 2011). Public–private partnerships in Africa have proliferated as a response to these conditions, converging around the agenda of a new Green Revolution for Africa (Moseley 2017; Amanor 2011; Scoones and Thompson 2011).

As Patel (2013: 4) articulates, "In pushing for a 'second Green Revolution', the first Green Revolution needs to be sold as a success." Because an argument for the success of the first Green Revolution was not possible in the African context (due to both maladapted agricultural prescriptions and practices that were not sustained), I argue that this narrative of the Green Revolution having missed Africa is critical for the rationale of a new Green Revolution agenda. That is to say that there are political benefits of framing these efforts to "transform" African agricultural systems as "new" and "uniquely"[10] promising, in that they obscure the role that some of the nGR's biggest proponents have played in creating the conditions for its supposed necessity.

The global food crisis of 2007–8 was a pivotal moment for the resonance of a framing of food security as a productivity problem and the push for a new Green Revolution for Africa. As Moseley (2017: 184) shows, it brought together "philanthropic foundations, industrialized country governments and aid agencies, agrifood companies, international agricultural research centers, and African governments ... in an unprecedented and highly coordinated effort to both frame African hunger as a supply-side problem and propose yield-enhancing technologies and market integration as the best way to address food insecurity in the region." The 2007–8 food crisis also revealed the lucrative potential of food commodity futures trading and attracted the world of finance to agricultural development, particularly on the African continent (Clapp 2012: 141; Roepstorff and Wiggins 2011: 24). This convergence of actors and the global food crisis led to the establishment of the New Alliance for Food Security and Nutrition. This initiative reflects a $3 billion commitment by the G7[11] countries as well as 21 African and 27 multinational companies[12] (Moseley, Schnurr, and Bezner Kerr 2015) and supports the accelerated implementation of the Comprehensive Africa Agriculture Development Programme, discussed in Chapter 6, via public–private partnerships.

The Alliance for a Green Revolution in Africa (AGRA) is a key implementing partner of the New Alliance for Food Security and Nutrition. Formed in 2006, AGRA was "born of a strategic partnership between the Bill & Melinda Gates Foundation and the Rockefeller Foundation to dramatically improve African agriculture, and to do so as rapidly as possible" (AGRA n.d.e). In Ghana, AGRA has supported plant breeding and commercial agriculture graduate programs, Ghanaian seed and agro-chemical company start-ups, and legislative change governing the agricultural sector—work that I discuss further in the chapters that follow. Proponents of the Green Revolution in Africa present increased access to hybrids and fertilizer alongside the introduction of "pro-poor" biotechnology as critical to addressing food insecurity and the impact of climate change. As noted in Chapter 2, proponents tend to present African agricultural systems through a crisis framework, the "tens of millions of people who are living on the brink of starvation" that require the intervention of a "uniquely Green Revolution" to "unleash the continent's agricultural potential."

As with the introduction of hybrids during the first Green Revolution, modernization of agricultural technologies remains a key focal point of the new Green Revolution for Africa. In both periods of agricultural change, efforts are centered on integrating farmers into the market economy through the introduction of hybrid or genetically modified seed. The adoption of "improved" (also known as commercial) seeds draws farmers into the market economy in two ways. First, these new technologies are designed to generate surplus that will be sold on the market, rather than meeting local food needs. Farmers' market integration has also been supported by a focus on elite, progressive, "serious" farmers that could demonstrate the successful use of these new technologies that then could be emulated by others, as I discuss in Chapter 5. Second, both hybrid and GM crops create dependencies and vulnerabilities, as both have mechanisms to prevent the saving of seeds and require intensive application of inputs to support them. The reduced fertility of hybrid seeds in subsequent replanting provides a strong incentive for farmers to purchase hybrid seeds each season. GM seeds are patent protected, which makes saving of these seeds a violation of contract and prohibited by law. Therefore, farmers that rely on either hybrid or GM seeds will need to purchase their seeds each season from agro-dealers (along with supporting inputs), rather than rely on seed saving. The dissemination of these two technologies reveals a continuity of focus on farmers' technology adoption as an important driver of agricultural change.

Furthermore, the introduction of GM seeds could not have occurred without the prior collection of germplasm from biodiverse regions achieved through the work of CGIAR during the first Green Revolution. The Green Revolution's Mexico program not only was about the introduction of dwarf hybrid wheat but also enabled the collection of maize germplasm from Mexico that advanced the development of hybrid maize in the United States (Kloppenburg 2004).[13] In other words, the collection of germplasm from the locus of maize biodiversity during the first Green Revolution created the conditions of possibility for subsequent breakthroughs in plant breeding (Brooks 2005: 362). Kloppenburg identifies that although there is extensive scholarship on the introduction of improved varieties to countries like Mexico and India, there is little attention on the transfer of genetic material from Mexico to the United States. This exchange of maize germplasm as a free good represents a common pattern whereby plant genetic resources from the developing world are conceived of as part of the "common heritage" that can be collected widely and freely and do not need to be remunerated for.

Genetic transfer through global germplasm collection, sometimes referred to as "plant hunting" programs, has value in the billions of dollars for its role in the development of agricultural commodities produced in advanced capitalist countries (Kloppenburg 2004: 157). The work of CGIAR-supported research in Africa also collects biodiverse local germplasm of cowpea, millet, and sorghum. Not unlike the first Green Revolution, international agricultural research centers utilize local germplasm for agricultural research and development of varieties such as drought-tolerant grains. Depending on how the products derived from this genetic material are handled, this could be accessed as a free good (as with the FAO "International Undertaking on Plant Genetic Resources"[14]) or as a commodity and intellectual property. I explore the expansion and implications of the commodification of plant genetic resources in Chapter 3, where I discuss biocapitalism.

In both Green Revolutions, development planners present agricultural transformation as an engine for economic growth that can alleviate poverty and promote security. In the first Green Revolution, US geopolitical interests were concerned with expanding the reach of new agricultural technologies into countries with large peasant farmer populations as a means to diminish the possibilities of peasant revolts connected to famine (Perkins 1997). Food insecurity was seen as holding the potential to drive people to communism. Thus, agricultural technology diffusion was perceived as a means to stop the spread of communism and a matter of US national security.

In the African Green Revolution, the US State Department now promotes biotechnology as one of its issue domains in West Africa.[15] Biotechnology is seen as key to reducing food insecurity both through the potential increase in household incomes by way of greater crop productivity (through the reduction of pest or drought-related crop loss) and through micronutrient enrichment of biotech crops to help address malnutrition. Importantly, the acceptance of biotechnology in African countries is also considered a matter of US national interest because of the benefits of increased market access for American businesses (108th US Congress 2003; Cooper 2008).[16] Whereas Patel (2013) makes the distinction between the original Green Revolution and the new Green Revolution as a shift from the geopolitical to the biopolitical, I would argue that the US State Department's involvement in its promotion reveals that geopolitics is still important but takes the form of a "corporate geopolitics" whereby the US government acts at the behest of American corporations to advance their interests. I further advance this point in subsequent chapters through examining the legal reforms that advance and protect investors' rights and intellectual property rights in Ghana. The promotion of both biotechnology and the nGR agenda, more broadly, is achieved through the strategic arrangement of public–private partnerships.

In sum, both programs for agricultural change frame hunger as a problem of production to be addressed through technological changes to agricultural practices (Stone and Glover 2017). The first Green Revolution and the new Green Revolution for Africa both signify periods marked by shifts away from reliance on the reproductive power of nature toward the productive power of industrial inputs (which I discuss more fully in Chapter 5) and created new conditions of possibility for market expansion. They both can be understood as geopolitical interventions, if we consider how American geopolitics today is influenced by corporate interests. The support for productivist agriculture by development planners and foundations is accompanied by pressures on states to create an enabling environment for agribusiness, which I turn to next.

V. The Enabling Environment for Agribusiness

Whereas there is widespread critique of the harm that structural adjustment brought on to the agricultural sector of African economies and a distancing from the language of adjustment by development planners (e.g., World Bank 2007), I argue that the logic of structural adjustment is still present within the discourse of the enabling environment.[17] While the World Bank, IMF, and USAID no longer embrace the language of structural adjustment, there is a fixation on promoting an enabling environment for investment. I came to understand this language as important because it was not only echoed in dozens of interviews, presentations, and workshops I conducted and was engaged in during my fieldwork in Ghana but was also notably present in the World Bank, USAID, AGRA, CAADP, New Alliance for Food Security and Nutrition, and the New Partnership for Africa's Development (NEPAD) documents that I have analyzed. This term was invoked uncritically and often without definition, which makes it ripe for discourse analysis to unveil assumptions, rationalities, and power relations embedded within this language (Escobar 1995; Sachs 2010). This section examines the discourse of the enabling environment by identifying its origins and revealing an ongoing understanding that it is developing countries' *internal* conditions that need to be changed, rather than recognizing the role of *external* policy change to promote the equitable and generative economic growth that the new Green Revolution for Africa policies appeal to.

In order to understand the origins of the discourse of the enabling environment, I reviewed World Bank documents that appeared in a search for the term "enabling environment."[18] Although the first appearance of the term enabling environment emerges in 1989, the precedent for use of the term is in a 1986 World Bank cosponsored conference "The Enabling Environment Conference: Effective Private Sector Contribution to Development in Sub-Saharan Africa" in Nairobi, Kenya, as well as a 1987 World Bank report, *Chile Adjustment and Recovery*. Emerging out of the conference proceedings was the Nairobi Statement, which affirmed that the

> necessary conditions for growth are the improvement of the national economies of Africa, deep reforms in the functioning of the public sector, as well as the provision of sufficient incentives to private initiatives. (McLean 1987: 4)[19]

In the 1987 World Bank report, "enable" is used three times and "environment" is used eleven times to refer exclusively to the political and economic context, such as: "better credit environment for future investment demand" (iv, 9), "improved agricultural environment" (14), and "strong and stable financial environment" (39). While the "international economic environment" (i, vi) is acknowledged, policy change is being presented as necessary to undertake at the national level— as opposed to change at the global, international, or bilateral levels—in order to attract investment. Although "enable" rather than "enabling" is used, this resonates strongly with the meaning of the term enabling environment used in later reports.

This term came in vogue in 1989, with three World Bank reports using the term extensively (World Bank 1989a, 1989c, 1989d). The report *Developing the Private Sector* (1989c) articulates that the private sector requires three main elements, the first of which is

> a supportive (or "enabling") business environment consisting of a stable macroeconomic setting, economic incentives that promote efficient resource allocation by the private sector, and laws and regulations that protect the public interest but do not necessarily interfere with private initiative. (vii)

The other two elements are (1) infrastructure and human resource development services and (2) a financial system that "provides the incentives and institutions needed to mobilize and allocate financial resources efficiently" (vii). The report links the creation of an enabling environment with structural adjustment by discussing the need to improve "the enabling environment through structural adjustment lending" (8) and advising that "reforms in the enabling environment should feature prominently in future adjustment loans" (15). In the World

Bank report *From Crisis to Sustainable Growth—Sub-Saharan Africa* (1989a: 4), the institution connects hunger in Africa to the need to create an "enabling environment," stressing the importance of economic growth in the agricultural sector:

> Africa must not only dramatically raise the level of domestic saving and investment, but also greatly improve productivity—by as much as 1 to 2 percent annually for labor and about 3 percent for land. This requires an *enabling environment*[20] of infrastructure services and incentives to foster efficient production and private initiative.

It further articulates that "an enabling environment for agriculture" requires privatization of input supply, processing, marketing, exporting, altering property rights to facilitate land titling, and deregulating price controls—in short, structural adjustment (World Bank 1989a: 8).

It is no surprise that a precursor of the term originates in 1987 to describe Chile, as Chile is often understood as a laboratory for the imposition of neoliberal economic reform (Klein 2007). As reflected in the reports that emerged from 1989 on, the discourse of the enabling environment is wedded to the logics and practices of structural adjustment. And that language is ever present today in the push for a new Green Revolution for Africa. In the impactful *World Development Report 2008: Agriculture for Development* that heralded a shift in attention to agricultural development on the African continent, the World Bank asserts:

> The nation state remains responsible for creating an enabling environment for the agriculture-for-development agenda, because only the state can establish the fundamental conditions for the private sector and civil society to thrive: macroeconomic stability, political stability, security, and the rule of law. … There is now general agreement that the state must invest in core public goods, such as agricultural R&D, rural roads, property rights, and the enforcement of rules and contracts … Beyond providing these core public goods, the state has to facilitate, coordinate, and regulate, although the degree of state activism in these roles is debated. (246–7)

In this agriculture-for-development agenda—of which the new Green Revolution for Africa is a central part—the state is relegated to being the enabler, not the leader. It is public–private partnerships that lead the charge.

The term "enabling environment" distributes agency and diffuses responsibility—there is both the "enabler" that creates the initial politico-economic conditions, but there is also the "environment" that fosters private-sector growth somewhat independently. The state is the initiator of reform but is also cautioned against too much "state activism." By representing the state as merely one "stakeholder" among an assemblage of actors, the state's active and broad powers to govern become reconfigured to a more passive, narrow emphasis on its role in "investment facilitation" (World Bank 2018: 3). This rather formulaic approach to development (that parallels the formulaic approach to agriculture promoted in both Green Revolutions) is an example of "rendering technical," whereby "experts tasked with improvement exclude the structure of political-economic relations from their diagnoses and prescriptions" (Li 2007: 7). Such language depoliticizes reforms that are, in fact, laden with power struggles that can result in the unequal distribution of benefits and risks, and obscures the role of external economic dynamics and trade policies of donor countries that undermine African agricultural development.

VI. Conclusion

I argue that this new Green Revolution for Africa agenda is wedded to an ahistoric productivist lens that views problems of African agriculture as fundamentally issues of production and ignores the past roles that promoters of this new Green Revolution played in undermining African agricultural development (McKeon 2015; Mkandawire and Soludo 1999). The narrative that struggling agrarian livelihoods in Africa reflect primarily problems of poor production ignores both the legacy of colonialism and the history of bad advice that have been disseminated to African governments as conditions for the receipt of World Bank or IMF loans. The latter is akin to what Mitchell (2002) describes in the *Rule of Experts* as USAID and other development planners leave their own policy impacts outside of their analysis and naturalize phenomena such as food insecurity in Africa. That is to say that structural adjustment in the form of trade liberalization and dramatic cutbacks to agricultural research and development spending have contributed significantly to the contemporary agricultural problems that Ghanaian farmers experience today. As discussed in the preface, trade liberalization that flooded Ghanaian markets with heavily subsidized rice coming from countries such as the United States forced a huge proportion of Ghanaian rice farmers to cease rice production due to an inability to compete (Mittal 2009). Furthermore, these structural changes not only exacerbated income inequality among farmers but also sharpened regional disparities in Ghana as northern Ghanaian farmers struggled to send their products to the south due to poor road infrastructure that worsened following structural adjustment–era spending cuts (Weissman 1990: 1625).

While cutbacks to agricultural research and development have been recognized as harmful to African economies—as reflected in the African Union's Maputo Declaration—the players that had advocated for this shrinkage of support for the agricultural sector retain positions of influence in shaping this new Green Revolution in Africa. The World Bank–supported Ghana Commercial Agricultural Project (GCAP) rice cultivation schemes are examples of "inclusive" agriculture and finance but reproduce uneven distributional effects not unlike the first Green Revolution through displacement of women's traditional foraging activities in the establishment of some of the GCAP plots in northern Ghana.[21] The United States' Feed the Future food security agenda in Ghana supports technology dissemination and increased market access for smallholder farmers with a focus on three crops: rice, maize, and soybean. However, this food security agenda neither impedes the flow of underpriced, heavily subsidized rice coming from the United States into Ghana nor acknowledges the historic displacing effects of trade liberalization on the Ghanaian rice economy in the regions it is intervening. Rather, it provides access to technical information and agricultural technologies that give sustained roles for US development practitioners and global agribusiness into the future (Mkandawire and Soludo 1999: 137).[22] That is to say that the Washington Consensus's advice (of which USAID is a part) helped to weaken the Ghanaian rice economy that these institutions are now positioned to support, thereby deepening dependency on Western technical expertise and obscuring its past role in creating the conditions for intervention.

In the *Agriculture for Development* report, the World Bank acknowledges the following: "Structural adjustment in the 1980s dismantled the elaborate system of public agencies that provided farmers with access to land, credit, insurance, inputs, and cooperative organizations" (World Bank 2007: 138). However, they write about structural adjustment as an inevitability, rather than misguided advice that they played a central role in disseminating: "Many reforms of the role of the state in agriculture were introduced as part of structural adjustment made inevitable by the debt crisis" (World Bank 2007: 43). What is left out is any recognition of the World Bank's role in promoting structural adjustment. Rather, they suggest that it was the inadequacies in the emergence of a private sector and state withdrawal "that was tentative at best" that can be attributed to the poor outcomes of structural adjustment (World Bank 2007: 138). The World Bank comes closest to acknowledging that the promotion of adjustment policies was a mistake by stating, "Contrary to expectation, the dismantling of parastatal agencies led to only limited entry of private providers, mostly in high-potential areas," but fails to clarify that this expectation was their own (World Bank 2007: 154). As Bromley (1995: 340) argues, "Adjustment by the North is not on the agenda."

Furthermore, the critical issue of missing, inadequate, and incomplete agricultural data in African countries poses severe limitations for empirically grounded policymaking (Wiemers 2015; Berry 1993). Although the World Bank acknowledges these problems, it does not stop them from making recommendations (Wiemers 2015: 109). The narrative of the "failure" of state interventions in agricultural development that serves as justification for SAPs conveniently ignores the fact that there are problems with missing and incomplete data that undermine claims to conclusive findings (ibid.). The new Green Revolution agenda ignores this history of mistakes, self-serving development advice, and failed attempts at "win-win" strategies. While there is no debate about the necessity for an increase in agricultural research and development spending in Africa in order to promote food security and alleviate poverty, ignoring this history normalizes reliance on philanthropy and the private sector to fill gaps in the provision of public goods.

Chapter 2

PHILANTHROCAPITALISM AND THE POLITICS OF PUBLIC–PRIVATE PARTNERSHIPS

I. Introduction

The year the Alliance for a Green Revolution in Africa was formed, statements by the former president of the Rockefeller Foundation, Judith Rodin, and the founding chair of the Alliance for a Green Revolution in Africa, Kofi Annan, suggested that with a new strategy and partnerships "to bring about a uniquely African Green Revolution," the "agricultural potential" of the African continent could be "unleash[ed]" (AGRA n.d.e). Interventions coordinated with foundations and their partners, as well as African leaders, scientists, and farmers, promise the alleviation of suffering of "tens of millions of people who are living on the brink of starvation in sub-Saharan Africa" (Rodin, as quoted by the Bill & Melinda Gates Foundation 2006). These statements from 2006 also illustrate the rise of philanthrocapitalist development, whereby philanthropists concerned with "impact investment" advance efforts to achieve a "win-win" of meeting humanitarian objectives while expanding business opportunities. Former UN secretary-general Annan speaks directly to AGRA's focus on entrepreneur-focused impact investment with his emphasis on interventions "where they will have the highest payoff," a characteristic of philanthrocapitalism and a central concern of this chapter (AGRA n.d.e).

Public–private partnerships (PPPs) have been the favored vehicle to support an "enabling environment" (discussed in Chapter 1) for investment in African agricultural development and have become a widely accepted institutional alternative to the public sector's provision of services and facilities and to complete privatization (Moseley 2017; Harvey 2005: 177). This chapter explores what is at stake in agricultural development objectives achieved by way of PPPs that promote philanthrocapitalism, a defining feature of the "new Green Revolution" (nGR) in Africa (Thompson 2014; Moseley 2017). I contend that we can speak of "philanthrocapitalist development" given the financial, ideological, and agricultural influence that philanthrocapitalism has had on development priorities. This chapter considers: *who benefits from philanthrocapitalist development in Africa?* I address this central question by highlighting how agricultural development practice has been influenced by the win-win strategizing of philanthrocapitalism and pressure from development planners to create an enabling environment for investment. I argue that the increasing prominence of philanthrocapitalist-led development in Africa raises important concerns about accountability, power, and democracy. Furthermore, its expanding role can also be understood as part of the growing governance of humanitarianism (Barnett 2011).

The chapter is organized as follows. First, I define what I mean by philanthrocapitalist development and reveal some of the implications of this phenomenon in Ghana. Next, I elaborate on how this development strategy raises concerns regarding accountability, distributional effects, and agricultural priority setting in countries that have been the recipients of such development interventions. In the fourth section, I illustrate the dynamics of philanthrocapitalist development by examining philanthrocapitalist PPPs promoting commercial soybean production and supplemental soy feeding in schools in Ghana. This case study provokes the question: To what ends is commercial soy in Ghana being promoted? What economies are bolstered by this example of philanthrocapitalist development?

II. *Philanthrocapitalist Development*

Philanthrocapitalism is a form of charitable giving guided by the logic of, and happening alongside, flows of transnational venture capital. This new form of charity calls upon philanthropy "to become more like the for-profit capital markets" and is particularly popular among those new philanthropists that have made their millions in finance (*Economist* 2006; Bishop and Green 2008). Philanthrocapitalist giving is not just a strategy among the new philanthropists but also includes some familiar faces in philanthropy. Actors like the Rockefeller Foundation are traditional philanthropic organizations that have advocated for new strategies such as "impact investing," described by the foundation as efforts to help "address social and/or environmental problems while also turning a profit" (Rockefeller Foundation n.d.a). The Global Impact Investment Network distinguishes impact investment by its intentionality: its aim is to generate social or environmental impact as well as the generation of financial return (GIIN n.d).

In an environment in which aid-based development has been heavily criticized for creating dependency and failing to substantially reduce poverty, a business approach to addressing development challenges has had significant appeal (e.g., Easterly 2007; Moyo 2010). Philanthrocapitalists critique existing philanthropy and aid as too shortsighted and insufficiently results-oriented. The idea is for philanthropists to behave more like investors—allocating money in such a way as to maximize "social returns" through the promotion of entrepreneurship (Bishop and Green 2008). As Bosworth (2011: 382) correctly points out, Bill Gates and company are "deeply invested in the philosophical presumption that material quantity ... can reliably generate social quality." This commitment to the quantifiable—what Bosworth calls "quantiphilia"—is characteristic of philanthrocapitalism (382). Philanthrocapitalism measures success through tabulation of "results-based" outcomes, linear progression over time, and a singular system of accounting (Thompson 2014). This is exemplified in the domain of agriculture with the fixation on yield as a measure of agricultural success (where alternate measures could include resource use efficiency or compatibility with local technology, knowledge, and context). This focus on measurable outcomes and legible effects means that philanthrocapitalist development is drawn to quick technological fixes rather than complex political remedies, apparent in the nGR's focus on access to improved seeds rather than on land reform to generate agricultural productivity.

Philanthrocapitalism recognizes the need for infrastructure in order to support its work, "the philanthropic equivalent of stock markets, investment banks, research houses, management consultants and so on" (*Economist* 2006). The Alliance for a Green Revolution in Africa, the G7 New Alliance for Food Security and Nutrition, and USAID-supported Agricultural Development and Value Chain Enhancement (ADVANCE) program all demonstrate this work in action: the establishment of linkages to finance, input suppliers, and research, with a centralized strategic management plan as well as the provision of guidance to support legislative reform that creates the enabling environment for the commercial seed sector and the leasing of land. "Venture philanthropists" favor partnerships, are hands-on, and are more likely to interact with states that favor neoliberal partnerships. Charitable giving becomes another form of investment, as venture philanthropy seeks to increase revenue as well as obtain reputational benefits (ibid.).

In conceptualizing philanthrocapitalist development, I build off of Linsey McGoey's (2014, 2015) and Carol Thompson's (2014) work on philanthrocapitalism as well as broader critiques of humanitarianism and aid (Barnett 2011; Chouliaraki 2013; Müller 2013; Easterly 2007; Banerjee and Duflo 2011). This book extends Barnett's (2011) argument by aligning philanthrocapitalist development as a part of the growing governance of humanitarianism. Barnett's book *Empire of Humanity: A History of Humanitarianism* clarifies why humanitarianism can be likened to empire:

> They involve long-distance rule by one people over another; they lack legitimacy because they rule without the blessing or participation of the people; and power radiates downward and for the purpose of advancing the empire's interest. ... Empires are branded as illegitimate because of their authoritarian qualities, but humanitarian governance is hardly a paragon of democratic rule. It is only over the last few decades that humanitarian governance has incorporated the views of local populations—and it is debatable how much energy humanitarians have put into these efforts or how receptive they are to redirection. (221)

Democratic accountability and transparency is not required of philanthropists. Given evidence of donor-driven tendencies and inattention to metrics beyond business performance as well as the scope of philanthrocapitalist development projects, this lack of accountability and transparency inherent to philanthrocapitalist PPPs is problematic. But as Barnett (2011: 221) points out, the legitimacy of philanthropy and aid is not dependent upon "deliberation, dialogue, or even consent."

Hands-on venture philanthropy, such as the Bill & Melinda Gates Foundation, has had a significant influence on public policy in the developing world (Spielman, Zaidi, and Flaherty 2011). This ability to shape policy is achieved in part given the substantial budgets philanthrocapitalist foundations have: between 2006 and 2009 the Bill & Melinda Gates Foundation (the world's largest foundation) had invested more than US $1.4 billion in agriculture; by comparison, the United Nations Food and Agriculture Organization's total budget for 2010–11 was $1 billion (McKeon 2014: 4). In 2016, 35 percent of the budgeted revenue for the Ghanaian Ministry of Food and Agriculture came from donor funds, money that is not only substantial but also more reliably released.[1] Additionally, philanthrocapitalist development gives space for increased donor control of agenda setting through limited-purpose and invitation-only grants (McGoey 2015: 98). This influence raises concerns given that this form of philanthropy can also lead to contradictions in the foundation's goals: for example, foundation money spent on supporting public health, on the one hand, and foundation money invested in corporations that cause illness and disease through pollution, on the other (Boseley 2007).

How donors view problems and priorities may not align with how governments and citizens view them, leading to projects that may not fully reflect recipients' interests but often also require recipient countries to support these initiatives (financially or through the work of creating an enabling environment) as part of the PPP framework (McGoey 2014). Furthermore, McGoey (2014: 122) argues that it is governments, rather than philanthrocapitalists, that are the risk-takers and that they intervene in ways to attract philanthrocapitalist investment:

> Philanthrocapitalists have helped to perpetuate a dubious belief: the idea that corporations and private entrepreneurs are subsidising gaps in development financing created by increasingly non-interventionist states. In reality, it is often governments subsidising the philanthrocapitalists.

This can be seen in the area of agricultural extension in Ghana. While it is widely recognized that the ratio of extension agents to farmers is inadequate, leaving many farmers outside of the reach of that agricultural assistance, some actors involved in the promotion of value chains "go about sourcing some public extension people privately to train their extension officers."[2] Such a practice further stretches thin the alarmingly limited resources of public extension.[3] While these actors promote interventions to integrate farmers into value chains and get recognition for their efforts to "transform agriculture" and alleviate poverty, it is, critically, the public sector that plays a supporting role through the training of private extension. Philanthrocapitalists obscure the role that government performs and, moreover, aid the narrative that government does not work by, in this case, diverting public extension agents from their already unreasonable extension responsibilities. Philanthrocapitalists reaffirm the neoliberal argument that government's dysfunction necessitates privatization by contributing to that dysfunction.

Philanthrocapitalists' emphasis on measurable outcomes and impact investing means that so-called wicked problems and root causes, such as addressing how to sustainably fund improvements to the system of agricultural extension, are likely to be brushed aside in pursuit of addressing problems that can have quick, legible, and quantifiable effects (Ramdas 2011: 395). This focus on impact investing can be seen in discussions around the need to shift away from rice as a major staple crop of development focus in Ghana; erratic rainfall, dams in desperate need of de-siltation and repair, and the poor competitiveness of Ghanaian rice were highlighted as reasons to no longer fund rice improvement as an agricultural development objective, despite its importance to Ghanaian diets.[4] Additionally, the emphasis on entrepreneurs as a part of the nGR's philanthrocapitalist partnerships is likely to generate uneven distributional effects as this strategy does not directly support vulnerable populations but rather makes the assumption that the use of the same value system that generates wealth for the few can reap broader benefits to address social disparities (Bosworth 2011: 383). McGoey (2015: 112) refers to this central assumption as the proud embodiment and effective repackaging of trickle-down economics.

Philanthrocapitalism, not unlike traditional philanthropy, is also a legitimating mechanism. In the use of the market to tackle social problems, philanthrocapitalism advances no criticism for the model of economic growth that allows for vast social disparities and accumulation of wealth among the few (Ramdas 2011: 393). Rather, it deflates criticism of the exorbitant wealth of monopolists like Bill Gates: the more he gives to charity, the more his wealth is legitimated (Morvaridi 2012: 1193). Furthermore, his identity as a successful entrepreneur further validates his project to support other entrepreneurs. Philanthrocapitalism is related to the emergence of global corporate social responsibility initiatives, triggered through a dynamic interplay between actors within civil society and business (e.g., Ruggie 2004). Innovative, successful campaigns by activists have highlighted unethical corporate practices, such as the use of deforestation, sweatshops, child labor, and support for brutal military regimes, and have affected change in these practices (Klein 2000; Haufler 2009; Nixon 2011). In a context in which so much value is ascribed to a brand, an easily identifiable corporate brand is vulnerable to rebranding by activists. Corporate social responsibility is a means to change public perceptions of corporations and has, in effect, changed corporate practices.

An example of this legitimating mechanism at play is global agribusiness's funding of the development of "pro-poor" biotechnology, discussed further in Chapters 3 and 4. Global anti-GMO campaigns have been successful at framing genetically modified (GM) crops as dangerous to health and social welfare, and commercialization of maize, sub-Saharan Africa's only commercially grown GM food crop, has faced fierce resistance in South Africa. Presenting GM crops as serving humanitarian ends is a powerful way to re-characterize the work of controversial agribusiness companies like Monsanto and Syngenta. Furthermore, the "donation" by Monsanto of the *Bt* gene, the $273 million that the Bill & Melinda Gates Foundation gave to agricultural development in Africa (Spielman, Zaidi, and Flaherty, 2011), or other such charitable acts wield a certain degree of political influence in subsequent relevant decision-making. Spielman, Zaidi, and Flaherty (2011) find that the scale at which the Bill & Melinda Gates Foundation gives to support development infrastructure across the domains of agriculture, public health, technology, and education has given the foundation significant political influence in discussions regarding development priorities within multilateral aid agencies. Part of this influence has to do with the large financial commitments of the foundation; however,

> the Bill and Melinda Gates Foundation's influence on the global agricultural development agenda has been more than proportional to its financial contribution during the past five years. … The Foundation's comparative advantage comes from its refreshingly creative perspectives, its willingness to take on a strong leadership role, its openness to partnership with other donors, and its ability to operate without the excess baggage of a large bureaucracy or conflicting constituent interests that hamper many multilateral and bilateral donors. (Spielman, Zaidi, and Flaherty 2011: 4–5)

The political influence gained through philanthropy and the preference to engage with foundations that are free from "the excess baggage of a large bureaucracy or conflicting constituent interests" further elevates the role of philanthrocapitalist foundations within policymaking, without the mechanisms of democratic accountability or oversight (Patel 2013: 10; Barnett 2011; McGoey 2015).

Within this nGR, the private sector, foundations, and bilateral aid play expanding roles in agenda setting and the provision of (purportedly) public goods. For example, recent community participation in the Savannah Agricultural Research Institute's interventions to improve cowpea did not involve soliciting much (if any) feedback on the priorities of women farmers before proceeding to partner with Monsanto and USAID in the development of transgenic cowpea.[5] Rather, the community participation that this philanthrocapitalist PPP practiced was in the form of biotechnology awareness seminars that shared biased information about the benefits of transgenic cowpea to villagers in neighboring communities. Conversely, projects supported by philanthrocapitalist partnerships may also be enthusiastically supported by local recipients, but funders may choose to terminate project funding and are not required to provide justifications for shifts in support or priorities. Faculty I spoke with at the West African Centre for Crop Improvement (WACCI) at the University of Ghana, Legon, expressed frustration regarding AGRA's decision to cut funding to WACCI, despite the institute's successful recruitment and graduation of numerous PhDs in plant breeding.[6]

Much of the recent wave of investment in African agriculture can be identified as philanthrocapitalist—agricultural biotechnology PPPs in Africa in particular are characteristic of this idea (Thompson 2014; Morvaridi 2012; Ignatova 2015). Philanthrocapitalism is reflected in the work of AGRA and the African Agricultural Technology Foundation (AATF) and is most explicit in the motto of the US-based NGO and the largest recipient of Bill & Melinda Gates Foundation funding, TechnoServe: "business solutions to poverty."[7] Furthermore, microfinance, support for legislative changes in the seed sector, and the donation of genetic material for crop research are specific expressions of this phenomenon. Microfinance provides opportunities for impact investing through access to small amounts of credit that support farming investments. This also deepens capital's reach into rural towns and villages and allows new markets to take root. The Alliance for a Green Revolution in Africa and the Ghana Commercial Agriculture Project's guidance for legislative changes in the seed and land arenas, respectively, are intended to help small farmers and also to expand the private seed sector. The AATF, supported by USAID and the Rockefeller Foundation, is a "new and unique public-private partnership" that is designed to assist in the access of agricultural technologies for smallholder African farmers. The AATF does so through the negotiation of royalty-free transfers of patented technologies from companies such as Monsanto and Syngenta (US Committee on Science House of Representatives 2003). The "donation" of a gene by Monsanto, negotiated by the AATF to develop African crops for food security, such as *maruca*-resistant cowpea, creates new markets and serves to normalize a Western-style patent regime. The following chapter goes into more detail about this philanthrocapitalist PPP. Next, I examine the politics of PPPs in broader terms.

III. Public–Private Partnerships: Accountability and Interests

Drawing upon Akintoye, Beck, and Hardcastle (2003: xix), I define public–private partnerships as the public sector provision of an enabling policy environment for the involvement of private companies, NGOs, foundations, and foreign aid agencies in the financing, design, construction, operation, and ownership of a public sector service or utility. The idea of PPPs fits within neoliberal discourse that promotes smaller government and a growing role for the private sector but also emerges in the wake of realizations that privatization has been "oversold" (Ostry, Loungani, and Furceri 2016; Kessides 2004). The state provides the enabling environment for private sector investment through neoliberal legislative changes and "good governance." PPPs account for about 15–20 percent of infrastructure investment in the developing world (IEG 2015: v).

The nGR in Africa's central premise is that technology and PPPs can bring about economic gains that alleviate poverty and promote food security across Africa. PPPs are assumed to be politically neutral and uniquely advantageous: first, they bring together the strengths of the public and private sectors as well as civil society (in particular foundations); second, they are believed to be pro-poor, in the sense of being better capable of delivering services to the poor and possessing greater sensitivity to the needs and priorities of aid recipients; and, third, they are a mitigating influence on capital's excesses. Morvaridi (2012: 1195) argues that the philanthrocapitalist PPPs central to the nGR have been framed by global governance institutions as pro-poor on the grounds that they both make agricultural products available where the private sector has little commercial interest and the funds from philanthrocapitalist PPPs are more capable of reaching the poor due to their capacity to bypass government bureaucracy. When PPPs are philanthrocapitalist, they are based on assumptions that entrepreneurship and the pursuit of win-win outcomes can bring about gains for both capital and the poor alike. For example, as I talk about in Chapter 5, agricultural development interventions that reach "the last mile user" are believed to be a means to enhance capabilities for smallholder farmers to scale up and increase their incomes, as well as to develop local agribusiness. It is also, importantly, a way for global seed and agro-chemical companies (e.g., Monsanto, Bayer, Syngenta, DuPont Pioneer) to penetrate frontier market spaces at a period in which their practices have faced global scrutiny, that is to say, at a time that necessitates better public relations work and a concerted effort to expand their customer base.

PPPs diffuse responsibility among agents, socializing risks taken by the private sector. Through such partnerships with the public sector, global agribusiness corporations gain greater access to African policymakers and legislative tools. This is what Clapp and Fuchs (2009) refer to as "corporate agrifood governance," whereby corporations play a key role in influencing the rules to regulate their own behavior. Furthermore, these partnerships have a dual legitimating function: the public component of these partnerships can mitigate negative perceptions about corporate activity; the private component can lend greater legitimacy to governments criticized as being inefficient or corrupt. Governments are able to gain from advances in research and development and additional financial support that the private sector brings to the table. Companies benefit from an investment climate where risk is distributed and local knowledge of the uses of plants is accessible.

Moseley (2017: 184–5) argues that PPPs in the nGR are characterized by high levels of coordination and collaboration with global agribusiness as African smallholders are more tightly integrated into global markets by way of value chains. With the state reliant on private sector's support of agriculture, the private sector has had the opportunity to exert a greater influence on agricultural priorities in countries such as Ghana. The lack of state capacity to substantially fund agricultural research "accounts for the donor-driven nature" of research in the region (Puplampu and Essegbey 2004: 279–80). One such agricultural issue that has become a priority as a result of PPPs is the introduction of GM crops in African countries. Muraguri's (2010) study of agricultural biotechnology PPPs highlights problematic power dynamics within these agbiotech partnerships, such as the pathbreaking partnership between the Kenya Agricultural Research Institute (KARI) and Monsanto to develop a virus-resistant sweet potato. She finds that these projects are mostly donor-led, time-bound, and often disconnected from the "end users"—that is, the farmers that will utilize these GM crops (298).[8] Yet the resources and experience provided through these partnerships can be important for building national research capacity, as it did with the Kenya Agricultural Research Institute. This need is particularly acute considering how structural adjustment budgetary cuts undermined the capacity of African agricultural research institutes, as discussed in Chapter 1. Muraguri's study of agricultural biotechnology PPPs is significant both because of Nairobi's position as the epicenter of the nGR in Africa and because of the leading role that KARI has provided in initiating some of the first transgenic field trials in

Africa that have raised some concerns regarding benefit sharing in the domain of pro-poor biotechnology (Thomson 2002). Data on the effects of this model of public service delivery on the poor are scarce, in part, because the metric by which the success and desirability of PPPs are evaluated by the World Bank—the largest producer of expert knowledge on development—is by their business performance (McKeon 2017: 389; IEG 2015; Goldman 2005).

PPPs—by diffusing responsibility while consolidating influence and bringing together actors with potentially divergent interests—raise issues of democratic accountability. In resource-poor political environments, securing donor funds may be a central concern. If PPP-led projects are donor-driven, as Puplampu and Essegbey (2004) and Muraguri (2010) find and my research suggests, these projects may be at odds with the needs and priorities of the targeted populations. This indicates that we should be skeptical of claims of the inherent pro-poor nature of PPPs. To further investigate the impact of PPPs, I now turn to philanthrocapitalist projects to develop the soybean market in northern Ghana.

IV. Cultivating New Markets: Philanthrocapitalist PPPs and Commercial Soy in Ghana

Philanthrocapitalist PPPs are shaping agricultural diets and futures in Ghana. Here, I briefly highlight how philanthrocapitalist PPPs "cultivate taste" for soy by examining the work of getting certified soybean seeds to farmers, supporting soy supplemental nutritional feeding in schools, and promoting soybean as feed for the poultry sector, and analyze the political-economic consequences of doing so. This case study provokes the question: To what ends is a commercial soy market being promoted?

PPPs have been central to the promotion of commercial soy in Ghana as a nutritional supplement, as a potentially lucrative cash crop, and as animal feed, and this support of a non-native legume reveals insights into the role of philanthrocapitalism in the nGR in Africa. The Global Food Security Strategy (GFSS) for Ghana is a collaborative document of US government agencies with input from the government of Ghana, private sector, civil society, and academia to promote food security and to "put Ghana on a pathway to self-reliance" (Feed the Future 2018: 4). The 2018 GFSS for Ghana acknowledges health and income disparities between Ghana's north and south and also echoes President Akufo-Addo's expressed commitment to move "Ghana Beyond Aid":

> To help Ghana bridge the north-south divide, strengthen its ability to manage and finance its own development journey, and position the country to transition from a recipient of U.S. development assistance, to a U.S. strategic trade partner, Ghana's GFSS Country Plan will ... continue to improve agriculture-led growth, resilience, and nutrition in the Northern, Upper East, and Upper West regions of northern Ghana, where poverty and nutrition statistics are poorest. ... Second, GFSS will make select investments targeting higher-value commercial

crops across Ghana to support agricultural sector transformation through trade acceleration to put Ghana on a pathway to self-reliance. (Feed the Future 2018: 4)

Part of this strategy entails supplemental nutrition in schools with a focus on soy, "a relatively new crop in Ghana, introduced in the north with the goal of reducing malnutrition," and support for farmers to integrate into soy value chains (ibid.).

In the beginning of my research I was puzzled by the emphasis on soybeans as one of the three crops that were focal points of the Obama administration's Feed the Future Ghana program. Maize and rice made sense as they have had a very long history in the region and are considered part of local diets, but why soy, which has neither of those characteristics? Years later, I got a response from a Ghanaian economic counsel working on biotechnology for the US Embassy, a young man in his late-twenties to early thirties, who corrected me when I stated that soy was not a part of Ghanaian diets. He stated that he grew up eating "Tom Brown" at school, a soybean-enriched food aid product that had been distributed by Catholic Relief Services in partnership with USAID. This was part of US Public Law 480, which provided schools with corn–soya mix and wheat–soya blend products in partnership with religious-based organizations and UNICEF.[9] Such distribution was limited to southern schools and hospitals, which is consistent with the accounts from my northern informants that soy was not generally a part of peoples' lives.

The Obama administration's food security strategy in Ghana has focused on soy-based nutrition in Ghanaian schools, as has a 2018 UNICEF food security program operating in Ghana. When I asked my research colleague working on the program what he thought about the use of soybean in their supplemental foods—given that soy is neither a commonplace part of peoples' diets in the north nor a traditional food—he agreed that it was problematic and that "it would be much better if they built off of existing practices." Groundnut, cowpea, and Bambara beans, for example, are legumes that can and do provide protein in local diets (as do inexpensive dried and smoked small river fish that form a rich base for stews) when meat is out of reach. Other informants working on food security in Accra as well as the Soybean Innovation Lab (described below) stressed that soy is an "inexpensive protein" and packs more protein than other crops. It also creates opportunities for Ghanaian farmers to grow soybeans—another area of intervention to cultivate a taste for soy.

The Agriculture Technology Transfer (ATT) Project (discussed at length in Chapter 5) is one of the implementing programs for the US Feed the Future Ghana initiative, with the philanthrocapitalist organization the International Fertilizer Development Center[10] winning the bid to run its implementation. ATT is intended to "improve the performance of Ghana's agricultural research and extension systems by creating a private sector-led agricultural technology transfer mechanism, linking research, extension services and producers to a market driven approach to technology development and dissemination." Such a technology transfer takes the form of a "taste" of the product in question: twenty thousand "starter packs" of seed (soybean, maize, and rice), fertilizer, and inoculant have been distributed for free on the basis of a "try it and see" model through a local seed agribusiness that cost-shares with the ATT project (70/30, respectively).[11] Videos of entrepreneurial model farmers using the products are played in villages in northern Ghana to encourage them to purchase certified seeds out of USAID-funded seed vans, a form of philanthropy for the private sector.

Soy has been endorsed by development planners not only as a nutritional protein supplement in school feeding programs but also as an integral component in new poultry feed formulas, giving farmers another potential market for their crop. It became clear that soy's advantages expanded beyond nutrition—and were at least equally about the potential to grow new markets—when I interviewed a lead researcher for the Soybean Innovation Lab (SIL), the Feed the Future Innovation Lab for Soybean Value Chain Research. When asked about the SIL's portfolio, the researcher talked almost exclusively about soy production for animal feed. When I sought further clarification about the percentage of the project's efforts on human nutrition as opposed to animal feed, he stated "soybean is a feed crop, that's just a fact"; however, then he elaborated that the SIL portfolio on human nutrition is much larger and that they get "ten times the calls about that component" of the SIL's work.[12]

A couple of informants knowledgeable about US efforts to promote commercial agriculture in Ghana expressed cynicism about the ability of soy production to take off in Ghana. The propensity of farmers to struggle with "soy handling" due to problems of storage and spoilage as well as the possibilities that soy products may not meet the protein- and oil-content demands of a global market for poultry feed raise the question of whether Ghanaian poultry farmers are likely to use imported poultry feed if the price is right and the quality is high. As more poultry farmers use soy-based feed, this will increase demand for soy, regardless of where it is sourced, as the US Department of Agriculture's and the American Soybean Association's[13] philanthrocapitalist AMPLIFIES (Assist in the Management of Poultry and Layer Industries with Feed Improvements and Efficiency Strategies) by WISHH (World Initiative for Soy in Human Health) initiative affirms:

> The AMPLIFIES Ghana Food for Progress project, with a budget of $15 million over five years, will improve feed and poultry production in Ghana, West Africa, a market that already imports U.S. soy. As feed production improves and local poultry production grows, so will demand for U.S. soy. (American Soybean Association 2018)

I was alerted to this philanthrocapitalist project, a cost-share arrangement between the US Foreign Agricultural Service and the American Soybean Association, among others, through an interview with USDA officials in Accra who were excited to highlight their efforts to support the growth of the Ghanaian poultry sector;[14] the WISHH homepage, by contrast, defines the initiative as "U.S. Soy's Catalyst in Emerging Markets." This initiative is emblematic of the philanthrocapitalist flavor of nGR interventions supported by American interests: support for Ghanaian agricultural development is contingent upon a win-win of American businesses also benefiting from these programs. Furthermore, USAID's programs in the soy sector not only help to build agribusiness and demand for soyfoods but also help to increase demand for American techno-scientific expertise, as I discuss later in the book.[15] That is to say that philanthrocapitalist projects are part of developing demand for American agribusiness products (e.g., GM and hybrid seeds, agrochemicals, animal feed, tractors, information and training modules, and later digital analysis and informatics[16]), even as they are also intended to address malnutrition and limited market opportunities for Ghanaian farmers.

V. Conclusion

When PPPs facilitate philanthrocapitalism, they are based on assumptions that entrepreneurship and the pursuit of win-win outcomes can bring about gains for both capital and the poor alike. Weakened agricultural institutions in Ghana have benefited from new resources made available to them through the work of PPPs, and, given their needs, donor and private investor support is increasingly attractive. However, PPPs are neither neutral arrangements nor pro-poor. They reflect a form of development that is conditioned on the win-win: that the gains of development are explicitly shared between the public and private sectors. This analysis suggests that philanthrocapitalism serves primarily capital and US corporate geopolitics, with philanthropy being a byproduct of the pursuit of powerful actors' interests. Philanthrocapitalism can aid in the pursuit of aims by governments, but given its donor-driven nature, it may serve donors far better than recipients.

Capitalist relations of agricultural production—owned by foreign countries—gain greater control over Ghana and the Global South because they are introduced, spread, and legitimized by philanthropic organizations.[17] The outsized influence that foreign development planners, working alongside philanthrocapitalist foundations, continue to have over agricultural development priorities on the continent necessitates critical assessment given their history in disseminating inappropriate and conflicted prescriptions that have undermined the resilience of African farmers (as discussed in Chapter 1). Soy is neither a major food crop in Ghana, nor does its cultivation and storage fall easily within widespread existing agrarian practices. Workshops on the cultivation, processing, handling, and storage of soybeans as well as trainings for use in school cafeterias are needed for soy's commercial success (Soybean Innovation Lab n.d.). As such, soybean's unfamiliarity in Ghana creates demand for foreign expertise. However, as a "taste" for soy is cultivated among poultry farmers and households, this also creates an entry point for global agribusiness's market expansion. While food insecurity and malnutrition in northern Ghana warrants attention, so do philanthrocapitalist development interventions that are *supply-led* rather than *demand-driven* and concerned more about future markets than current needs.

Chapter 3

BIOCAPITAL, "PRO-POOR" BIOTECHNOLOGY, AND LEGISLATIVE CHANGES IN THE SEED SECTOR

I. Introduction

The previous chapters argue that we should be skeptical about the claims of newness and aims of poverty alleviation of the "new Green Revolution for Africa" agenda. Yet I would also argue, as have others (Moseley, Schnurr, and Bezner Kerr 2015; Thompson 2014; Brooks 2005), that there is novelty within this period of agricultural transition. I contend that this new Green Revolution for Africa advances the proliferation of new forms of capital that integrate biotechnology with philanthropy to create new market value.[1] These shifts are engendered by laws that entail changed relationships to seed, whereby seed becomes a patentable material, with specific regimes of access and use. I explore these relationships and legislative change through an analysis of the politics of genetically modified (GM) seeds in Ghana, an African state critical for the wider acceptance of GM seeds in a context of contestation. Legal changes taking place in countries such as Ghana support private-sector expansion by securing an "enabling environment" to make agricultural investment more profitable and to facilitate technology transfer. Philanthropy in the form of donated genetic material for the development of "pro-poor" biotechnology alters conceptions of seed and lays the foundation for its commodification. Yet this commodification and genetic manipulation of seed is resisted, and in this resistance, the traditional importance of seed is also reaffirmed.

The cultivation of GM crops is seen as critical for the economic development of countries such as Ghana. African countries have been framed as "laggards" in need to "catch up" with the rest of the world by embracing biotechnology. This view is captured by the statement of the "father of the Green Revolution," Norman Borlaug, and Jimmy Carter: "Africa has already missed the industrial revolution and the tractor and fertilizer revolution … there is a risk it will miss the biotechnology revolution as well"' (Paarlberg 2008: x). However, biotechnology proponents have recognized Ghana's developments in biosafety and biotechnology as worthy of emulation, positioning the country as a model for the region. In a July 2015 editorial in the *Graphic Online*, Dr. Florence Wambugu celebrated Ghana's biosafety law as "an excellent model that other African countries can emulate. … Countries in the region and other African countries can learn from what Ghana has done."

The framing of food production in African countries as a system in crisis has made technologies like GM seeds that promise greater productivity increasingly attractive and more likely to be adopted. Yet activists' counter-frames of GM seeds as threatening motivate calls for outright bans of this new technology. A recasting of the technology is necessary for market expansion, and this is being achieved through three primary ways: first, by framing GM seeds as an urgent necessity (the focus of Chapter 4); second, by presenting GM crops such as *Bt* cowpea[2] as pro-poor; and, third, by presenting donated proprietary genes as a "gift." Philanthrocapital—the merging of the logic of venture capitalism with philanthropy—serves to normalize seeds and genes as commodities.

The chapter begins by discussing the relationship between philanthrocapital and biocapital, two related forms of capital that have emerged as key sources of market value in the "new Green Revolution" in Africa. I then highlight the role of agricultural biotechnology public–private partnerships (PPPs) in facilitating the expansion of these forms of capital through the case of the Cowpea Productivity Improvement Project in Ghana (often referred to as *Bt* cowpea). This project reveals some of the novelty of this so-called revolution: the role of pro-poor biotechnology as a mechanism to minimize contestation over GMOs, the prevalence of new sets of expertise and partnerships that have emerged alongside the introduction of GM crops, the way that the commodification of seed can be furthered through the concept of "donation," and the manner in which these PPPs obfuscate questions of ownership of biological resources while also being presented as pro-poor entities in and of themselves. I then examine how legislative changes in the seed sector promote these novel forms of capital accumulation.

II. Biocapital and Philanthrocapital: From Ecological Commons to Life as a Business Plan

Biocapital

I situate my discussion of biocapital—the capitalization of life itself—within the literature on the bioeconomy and biopolitics, which highlights how the life sciences and "enterprising nature" have become key areas of capital expansion and sites of power struggles (e.g., Sunder Rajan 2006, 2012; Jasanoff 2012; Cooper 2008; Helmreich 2008; Shiva 1997; Birch and Tyfield 2013; Birch 2017; Dempsey 2016; Foucault 1979; Brooks 2005; Rose 2001; Fairhead, Leach, and Scoones 2013). Jasanoff speaks of the role of law in creating biocapital or "lively capital" and asks, "When does something in nature become property?" She argues that what intellectual property law rewards "is the act of economic agency that takes something that was fixed, embedded, and immovable and makes it specific, dynamic, and commercially value-laden" (Jasanoff 2012: 156, 181). The isolation and transmutation of genes for the development of GM crops enacts the relationship of capitalization that Jasanoff describes. However, as living resources, seeds circulate in ways that are not "fixed" and "immovable." Yet I concur with

Jasanoff that the "commercially value-laden" element is considered the *Bt* (*Bacillus thuringiensis*) transgene that needs to be protected through intellectual property rights as well as through biosafety.[3] Jasanoff also suggests that in order for this capital to circulate, there needs to be fundamental cultural shifts in relationships to nature. In a similar vein, Aistara (2012: 128) shows "how the imposition of intellectual property rights to seeds is an attempt to 'remake' seeds. They move from naturally and socially co-evolved, genetically mixed, locally adapted, and freely reproducible cultural objects to genetically pure, individually created, legally protected, and globally tradeable products." These changes in cultural practices with regard to seed can be seen in the identification of seed as intellectual property (rather than a commons), in the domain of biosafety expertise, and in the promotion of "farming as a business," which I discuss in Chapter 5.

My work also adds to the literature on the rise of asset-based finance and financialization in the life sciences. Whereas Sunder Rajan and other scholars of bioeconomy place emphasis on the latent value of biological resources, Birch (2017) and Birch and Tyfield (2013) draw attention to the growing importance of asset-based economic processes (e.g., intellectual property rights, rent) and the role of the firm in creating value in the bioeconomy. Together, these approaches allow me to view GM seeds as an embodiment of biocapital that reflects traits that are simultaneously that of an asset and a commodity. GM seeds are a part of the expanding rentier-based economy since money circulates through the sale or donation of the licensing of patented technologies, not only through the sale of GM seeds (a material- or commodity-based economy) (Birch and Tyfield 2013: 3). As Birch (2017) and Birch and Tyfield (2013) discuss, value is socially constituted by political economic actors in the bioeconomy, rather than in the value of the physical product. As discussed in the previous chapter, the value of a firm is reinforced by the reputational benefits of partnering with foundations and governments to develop pro-poor biotechnologies, whether or not this purported public good will ever be realized (Sunder Rajan 2006).

Although the speculative economy enables future promises to generate value, I contend we must consider the materiality of GM seeds given struggles over control and access that arise from problems of cross-pollination, unlawful seed saving, or biopiracy that appropriates indigenous knowledge. Whereas the intellectual property rights related to GM seeds present these proprietary seeds as an asset, their commodity potential is also seen as critical to transforming the agricultural landscape of northern Ghana. Ghanaian and World Bank development experts identify northern Ghana as having low reliance on commercial seed, fertilizer, irrigation, and mechanization (World Bank n.d.b). In order for a "new Green Revolution" to take place, both assetization and commodification need to take deeper root in places such as northern Ghana.

The natural characteristics of seed serve as a biological barrier to its commodification, as seed is both the means of production and the product (Kloppenburg 2004: 10–11). The US-consolidated seed industry took steps to overcome these barriers to commodification through the breeding of hybrids and the development of plant biotechnology, as well as through legislative change. The initial steps to legally enclose the seed as a form of private property coincided with major developments in biotechnology. The lobbying of large agricultural interests, pharmaceutical, and genetic engineering industries for a global and uniform patent regime enabled the transformation of the seed into a commodity (Kloppenburg 2004: 323). Subsequent legal developments in intellectual property law, like the 1991 revision to the Union for the Protection of New Varieties of Plants, removed the farmer's privilege, making it possible to infringe upon the tradition of seed saving.

The interpretation of biological material as patentable was globalized in 1994 through the World Trade Organization's Trade-Related Aspects of Intellectual Property Rights Agreement. TRIPS can be conceived as a legal means to enclose life as the private property of individual innovators, which enabled the patenting of biological materials and required all members to protect intellectual property. GM seeds are no longer just biological material, but rather an informational, proprietary *technology* licensed to farmers and protected through legal means (Brooks 2005). The discursive effect of these patent laws is that they redefine life as a form of property, as a commodity that can be traded and an asset for investment. The representation of life in informational terms, as a commodity that can be bought and sold, is a way by which life itself has been turned into a business plan. Capital now becomes biocapital (Sunder Rajan 2006: 41).

Biotechnology development and its accompanying patent regimes reflect a form of commodification of knowledge or what Sunder Rajan identifies as "biocapital," whereby biological life becomes a key source of market and informational value. Sunder Rajan considers biocapital to be an integral component to a new phase of capitalism.[4] As the corporatization of the life sciences industry deepens, pharmaceutical and plant biotechnology companies have continued to rely on indigenous people to share traditional knowledge of the use of plants in order to develop new drugs and seeds. Bratspies (2007: 317) argues that one of the reasons indigenous communities have faced such an uphill battle in trying to gain recognition of their property and culture is based in the incompatibility of indigenous and Western individualistic notions of property. As Bratspies explains, "Most legal regimes award the mantle of 'property,' with its attendant rights, only to the tangible goods produced by indigenous cultures, paying no attention to the contexts in which those goods were produced and used" (ibid.). In a similar vein, Jasanoff (2012: 181, 168) argues that "private takings," that is, "taking things out of public nature for private gain," requires genes to be isolated and taken out of its natural context whereby "genes are no longer nature's instruments ... but are amenable instead to human intentions and purposes." Biocapital advances a view of biological life that locates value in the stock of genetic material that divorces the seed from its social and ecological context (Escobar 2008: 140).

That is, the formation of biocapital focuses on the gene[5] as the source of market value, rather than the socio-ecological context from which germplasm is reproduced. The plant genetic material extracted for use in genetic engineering is not merely the product of a laboratory, but of generations of farmer experimentation and plant breeding that typically go unacknowledged. The patented material is isolated from this basis in vernacular knowledge. Escobar refers to this as "genecentrism":

> And although biodiversity is seen as encompassing more than genes, the recognition of its genetic foundation suggests that it is in genes, not in the complex biological and cultural processes that account for particular biodiverse worlds, where ultimately "the key to the survival of life on earth" … is supposed to reside. (Escobar 2008: 140)

The dominant approach to biosafety also focuses on genes with its risk assessments that concentrate on the environmental and public health effects of possible gene flow. As Andrée (2005: 29–30) points out in his analysis of the Cartagena Protocol on Biosafety, the African Group's position that risk assessment should include socioeconomic impacts was not adopted in the final protocol. In Ghana, socioeconomic risks are seen to be outside the parameters of the risk assessment study but would rather be considered at the stage of public comment following field trials of GM crops.[6]

From this genecentric perspective, the proprietary material in a GM crop is the transgene, the gene construct that has genetic material from two different species. For example, this would be the insect resistance trait derived from the bacterium *Bacillus thuringiensis* introduced into the plant genetic material of the target crop through genetic modification. In the 2004 Supreme Court case *Monsanto Canada Inc. v Schmeiser*, the court defended the patent protection on the glyphosate-resistant gene introduced into Monsanto's Roundup Ready canola products. The defendant Percy Schmeiser, a Canadian farmer, was found guilty of patent infringement, despite his claim that he never purchased Roundup Ready seeds but rather the GM seeds had cross-pollinated with his canola plants. Schmeiser argued that Monsanto's claim of patent infringement was patent overreach and appealed to a previous ruling that had determined that higher order life forms could not be patented. The Supreme Court of Canada did not view this previous ruling as relevant and argued that patent law recognizes patent infringement when a significant part of the whole is subject to unauthorized use.[7]

In finding Schmeiser guilty of patent infringement, the Supreme Court of Canada refused to weigh in on questions of the moral desirability of genetic modification or on the possibility of unintended cross-pollination. Rather the presence of the gene and Schmeiser's decision to replant his seeds was designated as patent infringement. The court's decision in this case reflects genecentrism: whereas the ecological possibility of cross-pollination was not considered relevant, the use of Monsanto's genetic material was. Monsanto owned the canola plants on Schmeiser's farm because they expressed the transgene that Monsanto had developed. What this case demonstrates is the contradictory legal and regulatory friction over biocapital. Will patents apply to the whole plant, as reflected in the *Monsanto v Schmeiser* ruling, thus shutting the door to seed saving, sharing, and resale of these seeds?

Later, I show how regulation and valuation of biocapital is advanced as well as further complicated by the role of donation, facilitated by the work of the African Agricultural Technology Foundation (AATF), in the development of pro-poor biotechnology. I argue that the AATF does more than support the expansion of pro-poor biotechnology. It advances the conception and legal protection of biological life as patentable, ownable material to be protected through intellectual property rights. The work of the AATF represents a form of philanthrocapitalism, which I turn to next.

Philanthrocapital

As a novel form of market value creation, biocapital does not operate in isolation. Rather, it works in tandem with subtler forms of capital accumulation, such as philanthrocapital. Philanthrocapitalism, as discussed in the previous chapter, is a form of charitable giving guided by the logic of, and occurring alongside, flows of transnational venture capital. Yet, McGoey (2014) argues that philanthrocapitalists, however emerging from the world of venture capital, are not actually big risk-takers. Rather, they expect states in developing countries to pursue the legislative reform, the creation of an enabling environment of PPPs that distribute risk, and the promise of legible impact in order to attract this form of investment. Philanthrocapitalists recognize the need for such infrastructure and much of the support by the Alliance for a Green Revolution in Africa, the G7 New Alliance for Food Security and Nutrition, and USAID's Program for Biosafety Systems (PBS) is intended to create the market, legal, and distribution infrastructure to facilitate these investment flows. AGRA and the Ghana Commercial Agriculture Project's guidance for legislative changes in the seed and land arenas, respectively, are presented as assistance to help small farmers and to expand the private seed sector. The donation of a gene by Monsanto through the AATF to develop African crops for food security, as in the case of the *Bt* cowpea project, creates new markets and normalizes the Western-style patent regime advanced by biocapital.

Philanthrocapitalism is at the core of the work of the Alliance for a Green Revolution in Africa and the AATF. One of the key aims of the AATF is to facilitate the development of pro-poor biotechnology. Pro-poor biotechnology refers to the development of transgenic crops to suit the needs of the farmer and the diets of the local people. This concept is appealing in three ways: first, as a technology that appears to be appropriate for resource-poor farmers and is thereby inclusive; second, as a technology to be developed by and for Africans in their indigenous research institutions;[8] and third, as an example of a technological development motivated by humanitarian concerns, rather than profit. In this way, the introduction of this technology acknowledges concerns regarding the exclusionary effects of the first wave of "Green Revolution" technologies. Whether those crops identified as pro-poor will benefit their targeted demographic remains to be seen—most of these transgenic crops have not yet moved beyond the confined-field trial stage of development.

PPPs in agricultural biotechnology create opportunities for the expansion of biocapital and philanthrocapital. They enable actors within the partnership to gain greater access to resources, whether it be local germplasm or a proprietary transgene, that can expand biocapital. They obscure individual institutions' relative contributions and, in particular, the ways in which governments subsidize philanthrocapitalist projects that can reduce risks taken by philanthrocapitalists (McGoey 2014: 122). They also obfuscate questions of ownership of the end product of pro-poor biotechnology that can enable "private takings" of "public nature" (Jasanoff 2012: 181). Finally, the public and humanitarian components of these partnerships can mitigate negative perceptions about corporate activity, at the same time as the private component can lend greater legitimacy to governments criticized as being inefficient or corrupt. Through such partnerships with the public sector, global agribusiness corporations gain greater access to African policymakers, legislative tools, and public research.

In my analysis of the development of *maruca*-resistant *Bt* cowpea in Ghana at the Savannah Agricultural Research Institute (SARI), I discuss the relationship between the public and private sectors within the Cowpea Productivity Improvement PPP as well as the role of both legislative change and donation in simultaneously attracting private investment and mitigating concerns regarding GMOs in Africa. It is to the development of pro-poor *Bt* cowpea in Ghana that I now turn.

III. Bt *Cowpea and the Growing Reach of Philanthrocapitalism in Ghana*

The global resistance to GMOs makes the introduction of GM crops politically challenging. This resistance to GMOs is found across the African continent, particularly in South Africa, where *Bt* maize is cultivated at a commercial scale with seed and agro-chemicals provided by agribusiness corporations Monsanto (acquired by Bayer in 2018) and Syngenta. Proponents of GM crops have sought an image makeover to ease the diffusion. Efforts to push commercialization of GM crops in other African markets have not focused on GM *Bt* maize and glyphosate-resistant soybean, two of the most common GM crops in the world. Rather, the greatest research and commercialization efforts on the continent have been placed on the development of pro-poor biotechnology like *Bt* cotton or *Bt* cowpea.

Created in 2003 and supported by USAID and the Rockefeller Foundation, the AATF is a public–private partnership designed to assist in the technology transfer of agricultural technologies to African countries in order to develop pro-poor biotechnology. The AATF negotiates royalty-free transfers of technology for use during the crop development stage and has been active in supporting the development of the *maruca*-resistant cowpea and nitrogen-use efficient, water-use efficient, and salt tolerant (NEWEST) rice in Ghana. The AATF has also taken a leadership role in the Water Efficient Maize for Africa (WEMA) project, a PPP of the Bill & Melinda Gates Foundation; the Howard G. Buffet Foundation; USAID; the national agricultural research institutes in Kenya, Mozambique, South Africa,

Tanzania, and Uganda; the International Maize and Wheat Improvement Center; and Monsanto.[9] These projects are presented in a way that reflects a view of pro-poor biotechnology as a de facto public good: that is, it will yield benefits above and beyond what agribusiness corporations can capture by marketing them to African consumers. The *Bt* gene used to develop transgenic cowpea in Ghana is a result of the AATF's negotiations with Monsanto.[10]

Before GM cowpea could be tested, there needed to be legislative change. Following years of deliberation and the outreach of the Program for Biosafety Systems, the Ghanaian Parliament unanimously approved the Biosafety Act 831 in December 2011. This legislation enabled the approval of research on *Bt* cowpea. In Ghana, *Bt* cowpea trials were approved in October 2012 and planting commenced in September 2013.

The international agricultural research institute taking the lead on cowpea is the International Institute for Tropical Agriculture (IITA) in Ibadan, Nigeria, which has been researching cowpea resistance to *maruca*, a pod borer that can damage cowpea crops by 30–80 percent.[11] Notwithstanding advances in integrated pest management, such as the use of different times of planting and plant densities, the application of botanical insecticides like neem oil, and the use of mixed cropping systems proven to work synergistically to create effective resistance to *maruca*, over three decades of IITA research has failed to find a *maruca*-resistant variety of cowpea (Karungi et al. 2000; Adati, Tamò, Yusuf, and Hammond 2007). The difficulty in identifying a resistant variety through conventional means motivated the institution's pursuit of genetic modification of the cowpea. The national agricultural research institutes of Nigeria, Burkina Faso, and Ghana supply the research and genetic material of local cowpea germplasm. The AATF negotiated with Monsanto a royalty-free transfer of the Cry1AB (*Bt*) gene—the gene that expresses the desired insect resistance trait—to the project.

The role of national agricultural research institutes is critical for the development of pro-poor biotechnology. The confined field trial research for *Bt* cowpea in Ghana is taking place at SARI in Nyankpala, a village outside of Tamale, surrounded by fencing and 24-hour security. It is the site of field trials and farmer demonstration fields to educate local people about transgenic cotton and cowpea. As an official for the Program for Biosafety Systems in Ghana told me in May 2015, "If Monsanto was the one pushing it, I'm sure that farmers would be a bit hesitant. Because it's being pushed by their indigenous research institutions, through a government negotiation, that's easier to accept."[12] That is, philanthropic support that enables the national agricultural research institutes such as SARI to take the lead on research obscures the influence of the different politico-economic actors that are a part of these public–private partnerships.

Bt cowpea could offer benefits to farmers in places like northern Ghana because it addresses a pest that has had dramatic impacts on crop yields, is suited to the local tastes, nutritional needs, and local agroecological context, and is guided by scientists familiar with the concerns of local farmers. Yet this project also raises some difficult questions regarding the future distribution of benefits. The *Bt* cowpea project is an international collaboration, so it is unclear who exactly would be the direct beneficiaries of the commercial sale of *Bt* cowpea seeds. The question of patents, that is, who would hold the patented technology upon commercialization, as well as whether that patent would be exclusively for the transgene or for the entire plant, was an issue that many of the agricultural scientists and biosafety experts preferred to stay out of, were unclear on, or wanted to consider later.[13] The AATF has explained to me: "The *Bt* Cowpea plant variety will be owned by AATF in trust for the local partners."[14] The lead scientist of the *Bt* cowpea project understood that Monsanto would hold the patent. When I inquired about the Plant Breeders' Bill and patents on plants, the PBS's country coordinator stated, "I stay out of this" and referred me to the PBS senior advisor to Ghana. This advisor had an unclear answer regarding who would hold the patent in the case of the commercialization of *Bt* cowpea, though he recognized it as an important question. His initial response was that he was not really sure, that the rights would be with CSIR and SARI, granted on a royalty-free basis by the AATF for "humanitarian purpose." Farmers would be able to save seeds as this was "not a commercial thing" but rather a humanitarian gesture, as Monsanto has "donated" the gene royalty-free. When asked about patent rights and *Bt* cotton, he was much more clear: "That one is definitely patented."[15] This ambiguity suggests that we should anticipate contradictory legal and regulatory friction over biocapital in Ghana. Will patents apply to the donated genes? Will they apply to other transgenes (e.g., glyphosate resistance) in the same seed? Will the industry forgo the royalties on one patent, yet charge for another? Access to biocapital is contested and negotiable, and actors' ability to influence the terms of negotiation depends on their ability to wield power in the firm, legal arena, and the lab, rather than the farm.

The *Bt* cowpea project highlights four dimensions of novelty that characterize the new Green Revolution for Africa. First, AATF's work in persuading Monsanto to "donate" the *Bt* gene used to develop the *maruca*-resistant cowpea can be understood as an example of the growing reach of philanthrocapitalism. Donation offers reputational benefits to Monsanto, and this may help expand Monsanto's market reach in the future. The donation of proprietary material in the case of *Bt* cowpea helps to construct biocapital through normalizing the perception of the seed as patentable material and advancing Western notions of property. Second, the contestation over genetic modification has motivated strategies like these to change the perception of this new technology. The emergence of a discourse around pro-poor biotechnology has helped to change the reputation of GM crops by highlighting its potential role in food security in African countries. Third, as discussed in the previous chapter, PPPs, rather than the state, take the lead in the development of agricultural biotechnology but require an enabling environment that fosters the growth of the private sector. Fourth, this new technology operates within shifting legal and technocratic regimes that deem this technology both as property and as risk, which requires intellectual property and biosafety regimes (supported by state resources) in order to enforce property protections and manage these risks. Such processes thus depend on legislative changes in the African seed sector.

IV. Creating an Enabling Environment: Legislative Change in the African Seed Sector

We must revive and rebuild Africa's battered capacity for applied research and make research institutions a cornerstone of our efforts. This process should encourage a spirit of entrepreneurship and the incubation of private companies that commercialize innovations that come out of Africa's applied research centers at various universities. ... We must help Africans create legal certainty, predictability, transparency to help spur investment from the public sector. ... And we must act very quickly because technology is moving so fast, and if Africa is already behind and nothing is done, it is unbelievable what is going to happen in 10 or 15 years. We have seen Asia move, we have seen South America move. Africa is moving backwards.

This statement by Dr. Kilama, president of the Global Bioscience Development Institute, at a 2003 US congressional hearing on Plant Biotechnology Research and Development in Africa summarizes the rationale behind the push for legislative change and applied research to support biotechnology: to nurture an entrepreneurial spirit that commercializes innovation as a way to move Africa "forward" (US Committee on Science House of Representatives 2003: 48–9). During this hearing, Gordon Conway of the Rockefeller Foundation; Andrew Natsios of USAID; and Robert B. Horsch, vice president, Product and Technology Cooperation for Monsanto, testified to the need and business potential of agricultural biotechnology development in Africa. This is when the AATF, formed earlier that year, was introduced to Congress and eagerly received. With this turn of attention to the potential of agricultural biotechnology in Africa came a push for legislative change.

The G7 New Alliance for Food Security and Nutrition, the World Bank's Enabling the Business of Agriculture, USAID, and AGRA have specified commitments to supporting food- and agriculture-related policy change, with particular attention to reform of the Ghanaian seed sector.[16] One of the programs to achieve this goal is AGRA's Program for Africa's Seed Systems (PASS). This program includes policy and advocacy in legislation over seed and provides assistance for African seed company entrepreneurs. The seed laws promoted by PASS are a form of investors' protections for plant breeders. The Alliance for a Green Revolution in Africa's Policy and Advocacy Program for the seed sector worked with the Ministry of Food and Agriculture to influence seed and fertilizer legislation in Ghana, such as the Plants and Fertilizer Act, 2010. The World Bank describes Ghana's Plants and Fertilizer Act as opening "the door for an increased role for the private sector in producing seeds for a number of grains," which suggests an entry point for biocapital (World Bank 2012b: xi). AGRA's philanthrocapital has provided the most financial support for the Ghanaian seed sector, funding plant breeding in Ghana's National Research Institutes, providing postgraduate training for plant breeders, distributing grants for small private seed companies, and supporting the development of agro-dealers. USAID's Feed the Future initiative has also outlined means to support seed production and regulatory reform. The Program for Biosafety Systems, also led by USAID, has offered regulatory reform guidance for the Plant Breeders' Bill, as it had in the lead-up to the passage of the Ghana Biosafety Act, which allowed GMOs into the country.[17] The hotly debated Plant Breeders' Bill is an extension of the Plant and Fertilizer Act that AGRA's policy and advocacy work promoted.

The Plant Breeders' Bill (recently renamed the Plant Variety Protection Bill[18]) that has been debated in Ghana is in line with the policy commitments found in the New Alliance Cooperation Framework. Upon presidential assent of the bill, it would strengthen the rights of foreign and certain domestic plant breeders. As stipulated in Clause 23 of the Plant Breeders' Bill, subsequent Ghanaian legislation could not override these rights: "A plant breeder right shall be independent of any measure taken by the Republic to regulate within Ghana the production, certification and marketing of material of a variety or the importation or exportation of the material." The Coalition for Farmers' Rights and Advocacy Against GMOs, which includes Food Sovereignty Ghana, organized protests against the bill's ability to override Ghanaian legislation and its requirement that seeds be stable and uniform in order to apply these protections. The organizers are concerned that these requirements, while protecting foreign plant breeders capable of meeting them, may be too high for local plant breeders with limited capital and will thereby exclude them from protection. This can contribute to further erasure of local plant breeders' contributions to improvements in agriculture.

The overall thrust of these efforts to transform the Ghanaian seed sector is a part of a continent-wide move to harmonize seed laws in order to encourage investment. Whereas countries like Ghana have adopted legislation (such as Biosafety Act 831) that is more risk acceptant, the African Model Law on Biosafety adopts the precautionary principle. Activist groups like the South African–based African Centre for Biodiversity highlight that such harmonization of seed law could make it difficult for countries opposed to GMOs to prevent them from entering their country. For example, a uniform southern African seed law, via the Southern African Development Community, could assist the spread of GMOs, which southern African countries have been able to resist thus far in spite of South Africa's commercialization of GM crops (Thompson 2014: 400). In West Africa, Ghana is the furthest along in trying to pass a bill like the Plant Breeders' Bill. It is expected that if the bill becomes law, Nigeria and Burkina Faso will follow suit.[19]

V. Conclusion

In the face of this contestation, the promotion of pro-poor biotechnology in Africa has been deployed to reframe biotechnology as a humanitarian enterprise. Using the concept of philanthrocapitalism, I have demonstrated how the donation of such technology serves to advance new markets in Africa under the pretext of providing a philanthropic response to perceived food insecurity. This donation mechanism produces political effects: the gift of a proprietary gene by an agribusiness corporation for the development of pro-poor biotechnology reflects a conception of genes and seed as something that can be privately owned first and then given away. Philanthropy normalizes the seed both as a commodity and as an asset—rendering it biocapital—and a genecentric view of biological life that locates biodiversity in the stock of genetic material and divorces the seed from its broader social and ecological context.

The development of pro-poor cowpea not only advances the commodification of seed; it acts as a point of insertion for the agrarian transition to capitalist agriculture. It is unlikely that companies such as Monsanto will turn a profit with each of the pro-poor biotechnology products it invests in.[20] However, it can potentially gain market entry that can create conditions of possibility for marketing its other more lucrative products in African countries. Additionally, if successful, this effort will facilitate the penetration of the fertilizer industry, farm machinery, and agrifoods in subsequent stages that can reshape African food systems into capitalist relations.

I contend that the space created by neoliberal cuts to agricultural research and development spending has made PPPs and donation an attractive response to deficiencies in state funding as well as an instrument for the expansion of philanthrocapitalism. PPPs obfuscate questions of ownership and create better access to certain genetic resources (i.e., local germplasm held by national agricultural research institutions), while potentially restricting others through patents (i.e., patented GM commercial seed). A consequence of these pro-poor biotech philanthrocapitalist partnerships is not only improved access to transgenes from Monsanto or Syngenta by national agricultural research institutes in Africa but also the increased access to African resources gained by agribusiness corporations that make possible the colonization of existing breeding work for private profit.

In the case of *Bt* cowpea, Monsanto has the proprietary title to the gene that confers the desired trait, insect resistance. Within a genecentric policy environment, this gives Monsanto a potential claim to ownership. However, this transgene is but one component in the production of *Bt* cowpea as Nigeria's national agricultural research institution provided the cowpea germplasm. This germplasm was also the product of years of farmer experimentation that would not be recognized by a patent given such genecentric policy commitments. The composite structure of PPPs complicates questions of ownership and creates conditions for competing claims. For example, Monsanto has in the past attempted to enclose future research results developed through PPPs it has participated in as its property (Paarlberg 2001: 48).[21] Determining who owns what will determine who gets the patents, what the patents are for (one gene or the whole plant?), and therefore who ends up capturing the most profits from Africa's new Green Revolution. PPPs contribute to this evasiveness by lumping together so many different kinds of agencies that the question of who owns what becomes almost unanswerable. Consider the role that institutions like AATF, USAID, SARI, AGRA, and the Bill & Melinda Gates' Foundation have played in supporting pro-poor biotechnology: Who *owns* their contributions to getting *Bt* cowpeas out of the lab and into ordinary farmers' fields? How will the weight of separate contributions by Monsanto or SARI be evaluated when it comes time to consider issues of ownership? The question of ownership is not considered directly in the 2013 Plant Breeders' Bill.

This legal ambiguity and desire to settle these complicated legal definitions on seed patents at a later date may lead to unequal outcomes when it comes to sharing the benefits of GM crop commercialization. That is, those that possess the legal expertise to defend their investments in the development of *Bt* cowpea are unlikely to be the national agricultural research institutes; rather, under conditions of overlapping and competing claims to patents, large corporations such as Monsanto or Syngenta are advantaged by their ability to wield power and to hire expert legal teams to promote and defend their claims to ownership. One thing that is clear is that the idea of plants as patentable material, previously part of an excluded category in the Ghana Patents Act of 2003, is gaining some acceptance within policy circles. With this growing acceptance, the property rights regime in Ghana is shifting away from complex customary notions of property toward Western neoliberal ideas about ownership. Philanthrocapitalism facilitates this shift. This understanding of the commodity form of seed is further reinforced through specific legal regimes that recognize intellectual property and grant greater authority to those actors that possess biosafety and biotechnology expertise. Moreover, this agricultural model elevates the importance of the laboratory and the legal arena—rather than the farmer's field—to agricultural development.

Chapter 4

TECHNOLOGICAL SAVIOR OR TERMINATOR GENE? BIOTECHNOLOGY, FOOD SECURITY, AND THE POLITICAL ECONOMY OF HYPE

I. Introduction

The challenge of feeding a growing population during an age of declining arable land and natural resources has generated an intense debate over the adoption of new agricultural technologies, particularly genetically modified (GM) seeds. GM crops are framed alternately as a means to improve the lives of a growing population in the face of climate change and as undermining the capacity for life—in the form of a seed—to reproduce itself. As such, this genetic manipulation has been both heralded as a technology with great potential and condemned as a source of worldwide contention: the "Golden Rice" that will save children from blindness and early death or the "terminator seed" that will lead to the poisoning of our food supply and the collapse of biodiversity.

Despite the controversial nature of this technology, GM crop cultivation on the whole is rising: according to the International Service for the Acquisition of Agri-biotech Applications (ISAAA), between 1996 and 2013, the land devoted to GM crops increased hundredfold; by 2017, "up to 17 million farmers in 24 countries planted 189.8 million hectares … an increase of 3% … from 2016" (ISAAA 2013, 2017). However, GM crop cultivation is also geographically concentrated, as approximately 85 percent of this cultivation takes place within the United States, Brazil, Argentina, and Canada (ISAAA n.d.d). In order to understand the uptake of this technology, it is important to recognize not only the dualistic nature of these framings but also the commonalities between adversaries. This chapter explores how both proponents and opponents of the genetic modification of seeds deploy discourses of *emergency* and *salvation* to usher in, or to resist, biotechnology's reach in the developing world. Fervent proponents of GM seeds make inflated claims about the promise of biotechnology to address the plight of the starving both to encourage investment in new agricultural technologies and to mitigate fears. Ardent opponents of GM seeds, though motivated by genuine concerns about the socio-ecological implications of this technology, make a parallel move: they frame GM seeds as a catastrophe-in-waiting and use this sense of urgency to catalyze support. Thus, both poles of this debate collectively construct emergency.

What is at stake in the framing of GM crops as responding to, or creating the conditions for, emergency is the narrowing of the range of options considered for future food production. More specifically, I argue that this construction of emergency may foreclose deliberation over the full range of options on how to address global food insecurity, thereby favoring ready-made, "off-the-shelf" technical solutions controlled by experts. Both poles of the GM debate, in dramatizing future food scenarios, construct conditions whereby small steps, compromises, or piecemeal approaches appear deficient and inappropriate in the face of such urgency (Ignatova 2015). Because emergencies are conceived of as large in scale, responses are expected to match this scale; as such, global networks of trade and aid seem well suited for a task of this magnitude. This suggests that preoccupations with "feeding the world" can have the unintended consequence of disregarding incremental, inclusive, and endogenous approaches to food security that may be less resource-intensive and environmentally damaging.

In this chapter I outline how the dynamics of the GM crop debate construct dualities that generate emergencies of scale. The adoption of GM crops gets posed as an "all or nothing" economic game: the uptake of this technology opens a door to either economic prosperity or indebtedness. The literature on the GM debate emphasizes its polarized character but ignores the extent to which both the biotechnology industry and anti-GM activists employ similar tactics to shape public opinion and policy regarding genetically modified organisms (GMOs) (e.g., Herring 2010; Paarlberg 2008; Andrée 2007; Jasanoff 2005; Stone 2002a, 2010). Despite the divergent framings of how to address current and future agricultural challenges, both proponents and opponents of transgenic seeds employ a combination of hype and science in order to facilitate or impede the adoption of GM crops. I define hype as the deployment of simplistic, exaggerated claims to stimulate activity such as consumption, investment, and philanthropy or to mobilize activism. Hype in this context is a bid for attention that overstates and simplifies the benefits or detriments of GM seeds. An unanticipated consequence of this framing of GM seeds as risky is the subsequent demand for an industry of experts to monitor biosafety. An additional outcome of anti-GMO campaigns is the greater frequency of biotechnology outreach events, those efforts to educate, train, and mitigate against negative perceptions of biotechnology and GMOs that are organized by industry-influenced public–private partnerships.

In the following section, I discuss the way in which discourses of emergency and salvation circulate in conversations about global food security and delimit responses to food and agricultural challenges. Framing threats to future food production (like climate change, land pressures, and biodiversity loss) as an emergency necessitates an immediate response. Biotech proponents, which include industry, government officials, and other development planners, offer technological salvation from this state of emergency through the transgenic seed. GM seeds are presented as holding the potential to be adaptive to climatic change, allow for more intensified farming on existing arable land, and tackle nutritional deficiencies. Environmental anti-GMO activists push for bans and moratoria of GMOs in order to save the planet from the perceived threats to human and ecological health. Yet not all opponents of GMOs would be satisfied with merely a ban on this technology. For instance, the global food sovereignty movement differs from many mainstream environmental movements as food sovereignty constitutes a form of radical mobilization that appeals to larger aspirations of independence, democracy, and solidarity. Hype is revealed as a technique that may enable campaign mobilization and affect policy changes like a moratorium on GM crops, but that is unlikely to support farther-reaching goals of public participation and influence in agriculture and food policy design.

In the third section, I examine how science is used alongside hype in the debate over the cultivation of GM crops. Biotechnology outreach has emerged as a strategy by industry and related public–private partnerships to combat "GM myths" and improve the public's perception of biotechnology. However, it is not only the proponents of GM seeds that use science-based informational campaigns to influence public opinion and policy. Anti-GMO activists critique the science of genetic modification as reductionist and imprecise, and wage a parallel campaign that draws upon experts in toxicology, biology, and ecology to support their claims. Both opponents and proponents make use of hype as a bid for attention and utilize science as a bid for authority. This simultaneous generation of hype and science creates confusion: it becomes difficult to tease out credible scientific and experiential information from an oversupply of hype-filled, web-based content on GMOs. Furthermore, the use of science as a means to legitimate critique of GMOs has exclusionary effects, creating barriers to public participation in this debate over how food should be produced and consumed.

II. Emergency and Salvation: GM Seeds and the Political Economy of Hype

Emergency

After the world's population reached 7 billion in 2011, a wave of reports and commentary framed the global food production and distribution system as one in crisis (e.g., National Geographic 2011). Widespread food riots as a result of dramatic surges in the price for staple foods in 2007–8 and another famine in the Horn of Africa were highlighted to convey the urgency and necessity of dealing with issues of chronic food insecurity.[1] Companies like Monsanto and Syngenta, industry leaders in the production of GM and hybrid seeds, seized and exploited this sense of emergency. According to Monsanto's website, in order to "to keep up with population growth more food will have to be produced in the next 50 years as the past 10,000 years combined" (Monsanto Company n.d.b). Syngenta also calls upon this discourse of emergency: "To feed this growing population, farmers will need to achieve at least a 70 percent increase in food production by 2050. Achieving food security won't be easy considering the megatrends of growing population, greater affluence, and increasing urbanization" (Syngenta n.d.). Implicit in these statements is uncertainty: How will we be able to produce this food?

This discourse of emergency is also reproduced by GM opponents that frame the diffusion of GM seeds as leading to the poisoning and corporate takeover of the world's food supply. Dying bees, lab rats with huge cancerous tumors, and cross-species mutant fruits become the symbols of the food emergency to which activists respond. Claims that "Monsanto's GM seeds create a suicide economy ... suicidal for farmers ... suicidal for the poor who are deprived food ... suicidal at the level of the human species as we destroy the natural capital of seed, biodiversity, soil and water on which our biological survival depends" (Shiva 2009) clash with assertions that "GM foods are safe, healthy, and essential if we ever want to achieve decent living standards for the world's growing population. Misplaced moralizing about them is costing millions of lives in poor countries" (USDA 2012). That is to say that proponents and opponents of GM crops alike utilize framings that connect the technology to the conditions for life itself.

I identify three key characteristics of this discourse of emergency: (1) temporal compression, (2) scale, and (3) perceived threat. Identifying a situation as an emergency impacts the way in which the problem is processed: emergency connotes a large-scale threat that necessitates immediate action (e.g., Lipsky and Smith 1989). One of the consequences of conceptualizing food insecurity and hunger as an emergency is that solutions are "rendered technical," skirting complex and deeply rooted sociopolitical issues (Li 2007: 7–10). In discussing famine, Edkins (2000: 1) argues that famine

> has been removed from the realm of the ethical and the political and brought under the sway of experts and technologists of nutrition, food distribution, and development. Its position there, as an appropriate subject for expert knowledge, remains a political position, but one can lay claim to a political neutrality because of the specific way that science is construed as "truth" in modernity.

The significance of the categorization of food crises as emergencies is that this classification constrains alternative agricultural imaginaries by changing the speed of the response. As Calhoun (2004: 376) observes, "Emergency is thus a category that shapes the way we understand and respond to specific events, and the limits to what we think are possible actions and implications." If immediate action is a mandated response to emergency, then this eliminates the possibility of taking slow, incremental steps to addressing the identified threat.

This "evental" character obscures the visibility of the structural processes of capital accumulation that might contribute to exacerbating the likelihood of future emergencies. Unlike addressing patterns and processes that may require political action and social change, an isolated event is likely to be more responsive to technological interventions. Furthermore, global funding structures of development projects incentivize this framing of emergency; routine upkeep and project maintenance are rarely attractive needs to support. The prioritization and designation of emergencies within these funding structures can be considered ways of minimizing institutional commitments to address underlying structural issues. Within this logic of emergency, rich nations can afford to be very risk-averse, "but the vast majority of humankind does not have such a luxury, and certainly not the hungry victims of wars, natural disasters, and economic crises" (Borlaug and

Carter 2008: ix). In this way, the discourse of emergency forecloses meaningful debate regarding how food should be produced and distributed and may render additional scientific testing of GM seeds "unnecessary," thereby reducing the time that activists have to respond.

Salvation

The invocation of a food emergency discourse prompts a demand for solutions. How can we feed the world with climate change threatening to wreak havoc and population continuing to grow exponentially? How may humankind be protected from the threat of toxic GMOs and corporate domination? In this constructed condition of emergency, a discourse of salvation is likely to have greater resonance. This discourse works in tandem with the emergency discourse to facilitate or impede openings for the entry of GM seeds. Biotechnology is presented as a means to "save" "poor Africans" from hunger (Glover 2010a).

The discourse of salvation has three key characteristics: it is transformative, faith-based, and dependent on the identification of a population "in need." Salvationary discourse is transformative because it promises alleviation from the current condition of crisis and emergency. Claims about the "miracles of modern science" reflect this aspect of the discourse. The discourse of salvation deployed by proponents of GM seeds reveals great faith in the *promise* and *potential* of science to solve problems of food production. This faith in science is not contingent on proof. Salvation also requires a population in need: diverse livelihoods become converted into the "misery" of the underdeveloped, a monolithic population that needs to be "saved" through technological interventions (Escobar 2010: 150). Hunger and malnutrition are linked to accounts of the world's poor as "starved for science" (e.g., Paarlberg 2008).

I borrow this notion of salvationary discourse from Kaushik Sunder Rajan's book *Biocapital: The Constitution of Postgenomic Life*. Sunder Rajan finds that in the domain of genomic research the "therapeutic molecule" can be used to invoke the future possibility of life-saving treatment "which of course need never actually be realized, but ... whose existence as a future goal is vital to the dynamics of the present" (2006: 48). Sunder Rajan argues that supporting the promises of the life sciences industry is an underlying belief structure of salvation. The narrative of drug development is one of a "miraculous enterprise" whereby the drugs themselves are the instruments of salvation (Sunder Rajan 2006: 35, 186–7). The belief in the promise and potential of science to offer life-improving technologies in the domains of human genomic research and agricultural biotechnology is the driving force of this discourse of salvation.

In a parallel fashion, high-yield varieties of seeds during the "Green Revolution" have been portrayed as the miraculous means by which India avoided famine (e.g., Paarlberg 2008: 8). Proponents of GM seeds have drawn upon this narrative to promote a "new Green Revolution for Africa," most notably through the creation of the influential Alliance for a Green Revolution in Africa, funded by the Bill & Melinda Gates Foundation and the Rockefeller Foundation, as discussed earlier.

GM seeds and the technologies that support them (pesticides, fertilizers, irrigation, tractors) are seen as the way by which the poor can be saved from poverty.

Sunder Rajan (2006: 111) also argues that the expansion of the life sciences industry is driven by speculative capital: "Speculation and innovation both involve the articulation of *vision*. But it is articulation that takes a certain form, that of *hype*. Vision and hype are both types of discourse that look toward the future."[2] Hype is common in the information age; constant access to nearly boundless quantities of information encourages overstatements and inflationary claims to gain the attention of a busy and distracted audience. In order to generate support to fund anti-GMO campaigns, opponents may make hyperbolic claims about the impact of this new technology. Rather than presenting this technology as part of a trajectory of environmentally damaging technologies, GM seeds are presented as radically "new" in order to solicit donations and other forms of support. Hype, therefore, is not only a bid for attention but also a funding mechanism. As Mayke Zaag and Annelies Zoomers (2014: 7) argue in the context of global land grabs, hype is capable of mobilizing the money, power, and resources of investors as well as activists—whether or not any of the announced land deals have materialized or the land purchased has been utilized.

Both opponents and proponents of GM seeds rely on hype to raise capital and generate concern regarding this technology. This political economy of hype operates at two levels: first, hype is deployed as an attention-seeking mechanism to attract an audience, maintain relevance, or counter adversaries' claims; second, hype is used as a way to generate excitement and solicit funds both for future research and development and for advocacy campaigns (van Lente, Spitters, and Peine 2013; Zaag and Zoomers 2014). This use of hype to attract financial support is akin to what Anna Tsing (2005: 57) describes as the "economy of appearances":

> *Performance* here is simultaneously economic performance and dramatic performance. The "economy of appearances" I describe depends on the relevance of this pun; the self-conscious making of a spectacle is a necessary aid to gathering investment funds. The dependence on spectacle ... is a regular feature of the search for financial capital.[3]

Pioneering industries and campaigns are more inclined to use hype, or in Tsing's terms "spectacle," during critical times when support is most needed. Biotech proponents use hype at the research and development stage of new products or when products are criticized. Activists use hype during critical political economic shifts, like impending legislative changes on GMOs or prior to the introduction of new transgenic products into commercial markets. Hype can be an important tool in garnering the support necessary to successfully ban GMOs or, by contrast, in generating the capital for new expansions in biotechnology research.

Van Lente (2000: 43) underscores how technological promise stimulates action: "Statements about future technological performance are not received as factual descriptions to be verified or falsified in due course. Instead, they mobilise attention, guide efforts, and legitimate actions." An example of this is a 2009

Monsanto press release that states: "Monsanto is on the verge of a technology explosion." In this address to investors, Monsanto's CEO promised products with improved yield for growers on the horizon, concluding with the following "cautionary statement":

> Certain statements contained in this release are "forward-looking statements," such as statements concerning the company's anticipated financial results, current and future product performance, regulatory approvals, business and financial plans and other non-historical facts ... since these statements are based on factors that involve risks and uncertainties, *the company's actual performance and results may differ materially from those described or implied by such forward-looking statements.*[4]

It should be noted that such a disclosure is required by law and is common to all Monsanto correspondence to investors. Yet what is interesting about this is that investors know this and perpetuate this economy of hype. Speculative investment offers the possibility of huge returns when investors get in early on products that may later become successes. Expanded investment in agricultural technologies such as this can create a false understanding that these technologies are "tried and true" and constitute appropriate responses to food emergencies. For other examples of salvation and the political economy of hype, I turn to the two of the most publicized "success" stories, GM sweet potatoes and Golden Rice.

"Pro-Poor" Biotechnology: The Technological Savior

Sometimes the ones responsible for the perpetuation of hype are scientists themselves. Jennifer Thomson, the former chair of the African Agricultural Technology Foundation who has played advisory roles in the regulation of biotechnology in South Africa, argues that the media has been biased in its accounts of biotechnology. This bias has led to the neglect of the ways in which GM crops have "saved" people from hunger and malnutrition. According to Thomson, the media focuses their attention exclusively on biosafety fears: "We don't, however, often read headlines such as 'GM rice saves millions of Asian children from blindness' or 'GM sweet potatoes save African crop from virus plague'" (Thomson 2002: 3). What is interesting about these statements concerning GM rice and GM sweet potato is that Thomson is invoking the "therapeutic seed" *despite the fact the therapy has yet to be realized.* Neither GM rice nor GM sweet potato is commercially cultivated and therefore is not available for consumption (ISAAA 2018). Furthermore, after three years of field trials at the Kenya Agricultural Research Institute, results demonstrated that GM virus-resistant sweet potatoes were no less vulnerable than ordinary varieties, and sometimes their yield was lower (*New Scientist* 2004; Gathura 2004).

Another illustration of the interplay between the discourses of emergency and salvation is the use of Golden Rice to demonstrate the saving force and benevolence of biotechnology. In 2000, *TIME* magazine declared, "This rice could save a million kids a year" because of the transgenic crop's promise to address vitamin A deficiency (Nash/Zurich 2000). Two decades later, this transgenic vitamin A–enriched Golden Rice has been approved as recently as December 2019 (ISAAA 2019) and is still not commercially cultivated. The delay is attributed by biotech advocates as a result of the destruction of field trials and the slow development of a regulatory infrastructure in the wake of the controversy surrounding GM crops.

However, Stone (2015) highlights that Golden Rice fails to deliver on the hype. A combination of poor field performance—"yield drag," whereby Golden Rice has lower productivity than comparably varieties—and inadequate evidence that the "golden" trait actually has beneficial effects on the vitamin A status of vitamin A–deficient human beings is the real obstacle to its commercial availability (ibid.). Furthermore, questions arise as to whether it *is even necessary* given the success of conventional nutrition programs in its steady progress of decreasing vitamin A deficiency (VAD) levels in children (from childhood VAD levels of 40 percent of the population in 2003 to 15 percent in 2008) (Stone 2015: n.p.; Stone and Glover 2017: 88). However, activists that impede the development of micronutrient-enhanced GM foods are framed as selfishly imposing their food choices onto people who have unfulfilled dietary needs or in the extreme as having "the blood of … millions of children on their hands" (Moore 2014). A US House Representative at a congressional hearing on plant biotechnology in Africa likened the barring of GM crops as "border[ing] on genocide" (US Committee on Science House of Representatives 2003). Such claims reveal important slippages: the *potential* of a technology to address micronutrient deficiency is equated with *the cure* for blindness and *salvation* from early death.

The discursive effect produced is that agricultural biotech companies can position themselves as doing therapeutic, humanitarian work that de-emphasizes the profit-generating elements of their enterprise. Biotech proponents level charges against Greenpeace as having committed "crimes against humanity" for their campaign against Golden Rice and, in doing so, place the development of Golden Rice and other transgenic crops as significant technological contributions to humanity. This belief in the ability of biotechnology to improve the human experience may be genuine, and in the early years of plant biotechnology, many scientists shared the hope that such developments could address some of the agricultural and dietary needs of the developing world (Glover 2010a). Yet besides the promise of salvation that pro-poor biotechnology holds for small-scale farmers in the future, virtually all GM crops currently cultivated are those developed for industrial farming and dominated by a few transnational agribusiness corporations. The crops that are planted in the developing world are by and large insect-resistant *Bt* cotton and maize and herbicide-tolerant soy, little of which are grown for direct human consumption. Among those crops like Golden Rice that are "bio-fortified," there is still substantial work to be done to prove that reliance on one micronutrient-enhanced crop—as opposed to a diverse diet of nutritional, traditional foods—is the best way to improve nutrition and attain food security.

Biotechnology as "FrankenFoods"

Whereas proponents of GM seeds invoke the promise and potential of science to save poor Africans from hunger, opponents use hype of a different kind. Anti-GM activists have labeled GM foods "FrankenFoods" and compare the unwilling introduction of GMOs into the food system to being a "human lab rat" (Burke 2010). Activists like Vandana Shiva attributed GM seeds to farmers' suicides—hence the label "suicide seeds"—even before the seeds had been adopted in India. However, it is likely that these seeds may have exacerbated existing problems once they were adopted and have the capacity to deepen indebtedness among farmers (Stone 2010; Sridhar 2006; Desmond 2016). Anti-GM activists have perpetuated the idea of an infertile transgenic "terminator gene," although this technology has not been developed for commercial markets.[5] The film *Genetic Roulette* claims that the introduction of GMOs is "the most dangerous thing facing human beings in our generation." Activists used a groundbreaking study on the effects of Roundup-Ready Resistant maize feed on rats to claim that this was unequivocal proof that GM foods are hazardous to human health. Images of the laboratory rats with grotesque tumors went viral online and were used as a visual representation of the dangers of GMOs, despite the fact that the most robust findings were that of the negative health effects of the herbicide, rather than the genetic modifications of the plants themselves (Séralini et al. 2012).

What is visible is that both sides of this debate use hype but for different purposes. Industry encourages investment in technologies like new transgenic seeds through hype and projections that are, by their nature, uncertain. Advocacy organizations like Greenpeace that oppose GM crops utilize hype that invokes dystopian futures to catalyze support and attract donations. Ironically, this suggests that activists begin to act according to the rules of the speculative marketplace—both sides need to dramatize danger in order to attract funds, capital, and resources to operate.

III. Science Fights

In the global debate over the role of transgenic crops, the field experiment itself has become a political object: GM field trials have become the target of destruction in protests in countries like France, the Philippines, India, Australia, Germany, and Spain. Signaling major distrust both in state regulation to protect human and environmental health and in the intentions of multinational agribusiness corporations, destruction of these field trials has been frequently performed as a rejection of a "lab rat" status. In August 2013, 400 Filipino protestors destroyed field trials of Golden Rice near completion—a symbolic action that proponents viewed as tragic because of the potential of the vitamin A–enriched rice to combat blindness (McGrath 2013). Protestors, on the other hand, perceived the transgenic crop plantings as an unnecessary experiment with ominous consequences. Mark Lynas, who claims to be a former Greenpeace activist and had publicly apologized for his past leadership of anti-GMO campaigns in Europe, asserted in his report that such individuals were from the city and not farmers. He condemned the destruction of this "vital" research and questioned the legitimacy of the action (Lynas 2013). *New York Times* reporter Amy Harmon countered that the Philippine government and Golden Rice developers—whom Mark Lynas had relied upon in his report—had incentives to discredit these protestors (Harmon 2013).

These divergent accounts of responsibility for the vandalism of the Golden Rice field trial are reflective of the kind of competing claims to legitimacy found throughout the debate over GMOs. Biotech proponents such as Lynas frequently frame social movements in opposition to GM crops as "anti-science." Yet, in order to be taken seriously within these heightened global debates over food security, many anti-GM activists engage in "science fights," utilizing scientific knowledge to undergird their critique of GMOs. They employ scientific arguments that critique the reductionist nature of the science used to develop it as well as claims of farmers' deepening indebtedness as a result of reliance on costly seeds and inputs (e.g., Shiva 2000). The prompt dissemination of scientific research via transnational advocacy networks that demonstrates the harmful effects of transgenic crops has had policy ramifications and sparked greater industry scrutiny. It has led to, for example, France's regulatory decision to ban a variety of Monsanto's *Bt* maize as well as to the US Environmental Protection Agency's decision to require seed companies to submit data to the EPA about the toxicity of *Bt* maize pollen in butterflies or else lose the right to sell the product in the United States (Rosi-Marshall et al. 2007; Losey et al. 1999).

Incidences like these can be used to explain a pattern of rapid, aggressive critique of papers unfavorable to biotech crops within the scientific community. As Emily Waltz reports in *Nature*,

> Those who develop [GM] crops face the wrath of anti-biotech activists who vandalize field trials and send hate mail. But those who … suggest that biotech crops might have harmful environmental effects are learning to expect attacks of a different kind. These strikes are launched from within the scientific community and can sometimes be emotional and personal; heated rhetoric that dismisses papers and can even … accuse scientists of misconduct. (Waltz 2009: 27)

When a problematic paper comes out, pro-biotech scientists react quickly, criticize the work in public forums, write rebuttal letters, and send them to journal editors, policymakers, and funding agencies. Waltz does not find that the scientists' financial or professional ties to the biotech industry are the source of motivation for this forceful response. Rather, she states that many of them do feel that GM crops are safe and have great potential to deliver important societal benefits. Waltz mentions that many of the scientists that have been active in responding to these problematic studies have been researching transgenic crops since the late 1980s and some have been closely involved with the regulatory approval of the first GM crops. However, in the midst of these scientific controversies, industry and industry-supported scientists have become vocal participants in discussions scrutinizing scientific studies critical of GM crops.

The reactions to Emma Rosi-Marshall, a stream ecologist, and her colleagues' study on the effects of *Bt* maize on caddis flies provide an example of the vocal role of industry in critiquing scientific studies that identify harmful effects of GM crops. Rosi-Marshall and her colleagues found that the flies fed only *Bt* maize detritus grew at half the rate of flies fed conventional non-*Bt* maize detritus; their fatality rates were furthermore twice that of the caddis flies fed non-*Bt* pollen (Rosi-Marshall et al. 2007). While they were criticized for their experimental design, critics took greatest issue with the conclusion that the transgenic crops had unexpected ecosystem-scale effects. Following this study's publication in *Proceedings of the National Academy of Sciences*, Monsanto sent the EPA a six-page critical response to the publication and posted the letter online. Rosi-Marshall stated in an interview that attacks such as those made against her study appeared to be orchestrated. Accounts of researchers like Bruce Tabashnik reveal the plausibility of this claim. Tabashnik was warned by an etymologist, William Moar (who later took a position at Monsanto), that his yet-to-be-published paper— showing evidence of insect resistance to *Bt* cotton—would have "devastating" consequences and was warned not to publish it. Tabashnik published the paper, which was subsequently criticized by Moar at conferences and within the journal in which it was published (Waltz 2009). In a similar vein, the Séralini et al. (2012) study previously mentioned has been discredited by biotech proponents and subsequently retracted, arguably to serve political purposes. The study has since been republished by *Environmental Sciences Europe* (Séralini et al. 2014).

An additional challenge to academic research on GM crops is the patents that agribusiness corporations hold. Technology use agreements—industry contracts that establish the permitted use of patented seed—both forbid saving seed and restrict conducting research on it. Such practice is rationalized as necessary to protect intellectual property in a competitive marketplace. Although Monsanto began issuing academic research licenses that allow research on certain aspects of commercialized genetically engineered products such as agronomic and yield comparisons, oil seed content, interactions of introduced traits with the environment, the effects of GM feed, and research on pest management and resistance, there are important restrictions. Excluded from these agreements is research on breeding with plants produced from transgenic seeds, development of commercial and noncommercial methods for detecting the presence of patent-protected traits in seed, research on modifications or improvements to the patent-protected traits, and research on new products prior to their commercialization (Monsanto Corporation n.d.). So whereas these licenses signal a step toward greater transparency and the potential for improved research on certain aspects of GM crops, academic researchers will continue to struggle to understand the extent of cross-pollination of GM seeds with other crop varieties, the vigor of plants produced with GM seeds, and the potential for nontarget genetic alterations resulting from plant biotechnology. Furthermore, scientists have complained that research on GM seeds was ultimately up to company approval, as they would have to seek approval both to access the seed for research (negotiated on an onerous case-by-case basis) and to publish the findings (Stutz 2010).

The dynamics of these science fights reveal processes of exclusion at work at two levels: first, the need to use science in order for the critique to be considered legitimate, thereby creating barriers to entry on the basis of knowledge; and second, as a byproduct of patent protections that makes the free study, assessment, and regulation of GMOs difficult. Herring (2010: 85) highlights these processes in his analysis of the precautionary approach of the Cartagena Protocol on Biosafety that guides many countries' approach to biosafety:

> Cartagena logic created niches for salaried employees and consultants in global regulation, education, and testing activities. These are material consequences of framing; beneficiaries are sharply differentiated by class and cultural capital from average citizens.

These dynamics of exclusion create realms of inclusion for the proliferation of experts: expertise is needed not only to develop the products (agricultural research scientists and plant biotechnologists), determine their nutritional benefits (nutritionists), monitor their safety (biosafety experts), create the legal regimes to protect the intellectual property (legislators and administrators), set up programs to maximize economic benefits (development planners and economists), and demonstrate their effectiveness (expert farmers in demonstration fields) but also to create a guild of professional advocates that can use science speak in order to resist it.

Countering the "Myths": Information and Persuasion in the GM Debate

Actors on both sides of the GM debate use formalized rebuttals in the form of fact sheets, letters published in public online forums, and other publications to undermine the legitimacy of opponents' claims. Activist organizations such as Greenpeace, Friends of the Earth, and the African Centre for Biodiversity produce fact sheets and videos that aim to counter GM "myths." Food Sovereignty Ghana, in drawing attention to the Plant Breeders' Bill as an entry point for the increased presence of GMOs in Ghana, organized a capacity-building workshop to explain how this impending legislation may negatively affect "the interests of millions of smallholder farmers in Ghana" (Food Sovereignty Ghana n.d.). The African Centre for Biodiversity, a South African–based advocacy NGO, generates documents and presentations that critique the science of genetic engineering using detailed technical information about the processes of genetic modification. These are circulated to like-minded organizations across the continent in order to attempt to block the approval of new transgenic varieties. The pan-African organization African Biodiversity Network supports advocacy workshops of local organizations that not only provide the public with information about the socio-ethical issues but also criticize the imprecision of genetic modification on scientific grounds. Studies of the impact of GM crops and feed on rats, butterflies, and other insects have also been used for anti-GM and labeling advocacy. Scientists believed to be pro-biotech have been identified as "myth-makers" online as activists challenge the reputation and legitimacy of work of these scientists.[6]

A common narrative of biotech proponents in Ghana is that the "anti" groups— often referencing groups such as Greenpeace or Friends of the Earth—have been very influential in promoting myths and fear about GMOs. This made drafting a biosafety bill a drawn-out process. There was a need to counter the "spread of the gospel of anti," and one mechanism to do so was through biotechnology outreach.[7] One of the most prominent entities on the African continent that conducts biotechnology outreach, the Open Forum on Agricultural Biotechnology in Africa (OFAB), is intended to facilitate the "the flow of information from the scientific community to policy makers and the general public" (OFAB n.d.). Biotechnology outreach campaigns such as OFAB have identified GM crop "awareness creation" for farmers and the public at large as critical to the widespread adoption of this technology and have sought to correct the myths about GMOs. Biotechnology outreach entails the advocacy regarding the benefits of biotechnology among African legislators, media, academics, farmers, traditional leaders, and even high school students.[8] There is recognition that in order for GM crops to be widely cultivated, farmers and the public at large have to be willing to participate in the cultivation and purchase of such crops.

The keynote address of the director general of the Council for Scientific and Industrial Research Ghana at the "sensitization seminar" for *Bt* cowpea and agricultural biotechnology iterates the key role of biotechnology outreach as a necessary response to the purported misinformation of powerful oppositional groups:

> The fear that surrounds GMOs all over the world including Ghana calls for a forum such as this with the aim of educating the relevant stakeholders and the public ... not forgetting the heavily funded anti-GMO groups which are misleading the public with very little evidence and in ignorance.[9]

"Sensitization seminar" is a term used by actors within institutions that include the World Bank to describe communication strategies designed to make local communities "sensitive" to upcoming changes that may impact them. Sensitization seminars have also been developed around such issues as land registration and sale and gender inequality. These seminars are led by experts in plant biotechnology and biosafety and are designed for a public audience with the purpose of both informing local communities about an issue and creating a space for the public to ask questions. As a requirement under the 2011 Ghana Biosafety Act, communities and relevant stakeholders adjacent to the project were invited to attend an informational seminar on *Bt* cowpea hosted by the Savannah Agricultural Research Institute (SARI), where some of the confined field trials have commenced. However, the seminar's central goal, which involved over two hundred participants, including the regional director of the Ministry of Food and Agriculture, the district assembly, the local chief, farmers, scientists, a representative of the National Biosafety Committee, and a representative of the African Agricultural Technology Foundation, was to rationalize why this new technology is necessary and beneficial for the local farming community.[10] As such, the *Bt* cowpea sensitization seminar in Ghana is an example of biotechnology outreach and the discursive positioning of GM crops as a pro-poor technology (Glover 2010a).

In addition to awareness creation and advocacy, biotechnology outreach programs such as the USAID-supported Program for Biosafety Systems train scientists and bureaucrats in biotechnology stewardship. The language of "biotechnology stewardship" is a favorite of industry, implying that problematic outcomes of the technology are due to a lack of care, rather than inherent problems of the technology itself. Biotechnology outreach programs and biosafety advising by experts have served to promote the adoption of GM crops in the developing world, as has the shared experiences of farmers at international and domestic field visits. Industry has shown enthusiasm for donating transgenes to facilitate the development of pro-poor technology in the developing world and given commitments to "product stewardship."[11] Biotechnology outreach not only mitigates fears through awareness creation of the benefits of GM seeds but also improves the reputation of maligned companies like Monsanto. Ultimately, the greater exposure to these new technologies and improved public relations efforts enable multinational agribusiness companies to expand their market reach in the developing world.

IV. Conclusion

I argue that despite the polarized nature of the debate, both proponents and opponents of GM crops rely on a combination of hype and science in order to advance their positions. Both lay claims that either utilizing or banning GM seeds will contribute to environmental sustainability and improve livelihoods. These advocacy efforts to persuade publics of the detriments or benefits of GM seeds could not be continued without steady sources of funding. How then is this funding secured over the long term? One way to attract support is to frame the issue as one that has significant and long-standing implications, a matter of life and death. Whereas anti-GM movements frame transgenic seeds as threatening both the livelihoods of farmers and the fundamental building blocks of life through indebtedness and "biopollution," biotech proponents see tragedy in the resistance to technologies that could fight hunger and save children from blindness. Advocates on both sides of this debate can feel that their financial support— whether in the form of donations to anti-GMO organizations or investment in biotech companies producing drought-resistant seed—is critical in order to save vulnerable populations. The use of scientific research to substantiate certain claims or debunk myths is exercised to bolster the legitimacy of certain prescriptions or prohibitions disseminated by these advocacy groups. Yet these prescriptions and prohibitions intended to improve the lives of target populations are closely tied to donor funding priorities and fiscal cycles, and can often be at odds with the needs of those whose lives they intend to improve.

The benefit that proponents and opponents gain from the construction of emergency incentivizes the use of hype. This, in turn, produces important effects on the character and quality of discussions on global food security. Hype has affected debates over the future of food production in two primary ways. First, the rendering of GM seed technology as a special technology with a unique set of risks has created a demand for a particular set of experts to manage these risks. The consequent shift in authority to the domain of experts is likely counter to the objectives of food sovereignty and anti-GMO movements that aspire for greater democratic control over decisions related to food. Second, under conditions of emergency, simplistic, total solutions that follow a prescribed formula may be preferred over incremental, experimental, and procedural responses. In this sense, bans of GM seeds achieved through court injunctions can be an outcome of this emergency framing, and this may be a desirable end for some anti-GMO opponents. Yet for those opponents who view GM seeds as a larger symbol for deeply rooted problems of inequality, power, and injustice, the emergency framing—through its consequent shifts in authority and its narrowing of focus—may hinder efforts to address these social problems. In this way, the emergency discourse and its corresponding relief plans may do violence to contingency and the endogenous solutions of "mixers," and render the actions of individual farmers and communities insufficient in the face of such urgency. The next chapter looks at how the hype that promotes a shift from "from farming as a way of life" to "farming as a business" in Ghana may contribute to greater rural inequity.

Chapter 5

EXPERTS, ENTREPRENEURS, AND THE "LAST MILE USER"

I. Introduction

The World Bank's 2007 *Agriculture for Development* report, influential in shaping "new Green Revolution for Africa" interventions, opens with the following oversimplification of African poverty:

> An African woman bent under the sun, weeding sorghum in an arid field with a hoe, a child strapped on her back—a vivid image of rural poverty. For her large family and millions like her, the meager bounty of subsistence farming is the only chance to survive. (World Bank 2007: 1)

The African woman, generalized here as the "vivid image of rural poverty," is only managing "the meager bounty" with her hoe, weeding by hand in this arid landscape. *But there are other options*, the next sentence articulates: contract farming, laboring on large farms, petty trading of processed foods. Here, escaping poverty is a matter of choice.

This "image of rural poverty" juxtaposed to the "others, women and men, [that] have pursued different options to escape poverty" (World Bank 2007: 1) resonates with how aid agencies, agribusiness, and the Ghanaian government have represented farming communities in northern Ghana. Proponents of agricultural modernization programs in Ghana identify northern Ghana as requiring development interventions in order to increase commercial agricultural opportunities and alleviate poverty. Such interventions are framed both by the Ghanaian government as a means to rectify historic neglect of northern Ghana (e.g., SADA 2010) and by funders as "socially inclusive commercial agriculture" (e.g., World Bank 2012c). The wide expanses of land of the Northern Region promise a future of commercial agricultural success yet to be realized[1] as commercial agriculture interventions aim to "[transform] agriculture from a low-productivity subsistence-based sector to one characterized by high-productivity, integrated value chains, and extensive value addition" (Ministry of Food and Agriculture n.d.).

Between 2009 and 2019, domestic and foreign development planners in Ghana prioritized the idea of "farming as a business"—a focal point of this chapter's analysis—on the basis that such a strategy could benefit Ghanaian agribusiness and alleviate poverty simultaneously. The Akufo-Addo administration's Planting for Food and Jobs program, the World Bank–funded Ghana Commercial Agriculture Project (GCAP), the Mahama administration's Savannah Accelerated Development Authority (SADA) initiative, as well as the Atta Mills administration's Plant and Fertilizer Act of 2010 and Ghana Biosafety Act of 2011 are examples of efforts to modernize Ghanaian agriculture in line with the new Green Revolution in Africa agenda. Bilateral aid (in particular USAID), the Alliance for a Green Revolution in Africa, and agribusiness actors (from Ghanaian seed producers and agro-chemical distributors to Monsanto and Syngenta) have partnered with the government of Ghana (via the Ministry of Food and Agriculture; Ministry of Environment, Science, Technology, and Innovation; the National Biosafety Authority; and the National Lands Commission) and agricultural research institutions and universities (SARI, CRI-CSIR, University of Ghana, Legon, KNUST, and UDS Nyankpala) to promote "farming as a business." Different configurations of the aforementioned actors converge in public–private partnerships (PPPs) to tackle a range of aspects of this transition: the advancement of agribusiness entrepreneurship and the integration of farmers into outgrower contracts, the registration of land and the promotion of Western-style intellectual property rights, and the introduction of modern agricultural technologies such as hybrid and genetically modified seeds, fertilizers, tractors, and processors via private sector initiatives and "complementary and targeted public support" (Grow Africa Secretariat 2013: 42).

This chapter focuses on the nexus of relations between experts, entrepreneurs, and the last mile user—the farmer that "pursues different options" and adopts modern agricultural technologies and integrates into value chains—that are central figures in this transition from farming as a way of life to a business. I argue that the new Green Revolution for Africa interventions normalize a particular modality of farming that *may introduce new vulnerabilities* (e.g., expensive input packages whose formulaic nature does not respond well to deviation, reliance on irrigation systems that may not function, and debt) *as it mitigates others* (e.g., securing contracts, attracting investment, and making technologies more accessible). Putting my fieldwork in conversation with Amanor (2011, 2009), Chambers (1983), Scoones and Thompson (2011), Berry (1993), Richards (1985, 1993), Bernstein (2010), Shiva (1991), Scott (1976, 1998), Yapa (1996a), Escobar (1995), Mitchell (2002), Li (2007), Ferguson (1994), Oya (2012), Nyantakyi-Frimpong and Bezner Kerr (2017), and Yaro, Teye, and Torvikey (2017), I highlight the ways in which the promotion of a particular model of agrarian entrepreneurship is likely to benefit elite[2] farmers and demand for foreign agricultural products (patented seed, inputs, tractors and tools, education and training, development assistance and expertise), even as it is "building capacity" and "improv[ing] the enabling environment for farmers" (World Bank 2012c: n.p.).

The chapter is organized as follows. The next section of the chapter draws upon my fieldwork studying farming practices and agricultural development interventions in the Northern and Upper East Regions of Ghana in order to elucidate what is meant by making farming in Ghana a business rather than a way of life. It highlights three dimensions: the shift from the "'reproductive power' of nature" to the "'productive power' of industrial inputs" (Yapa 1996a), the professionalization of farming, and the promotion of entrepreneurship and agricultural risk-taking. The third section analyzes the economics, technologies, and assumptions of the last mile user approach to transform farming into a business and develop a commercial seed sector in northern Ghana. Next, I discuss the logic of agricultural "exit" and how this logic invokes struggles over identity: who should farm, who is an expert, who is an authority, and whose knowledge counts. I conclude with an assessment of whether the promotion of farming as a business is genuinely "pro-poor" and "socially inclusive" or another form of hype.

II. From Farming as a Way of Life to Farming as a Business

In an address to the FAO in Rome on June 27, 2011, former Ghanaian Minister of Food and Agriculture Kwesi Ahwoi characterized Ghana as being at a crossroads:

> Even though Ghana has achieved the [UN Millennium Development Goal 1], we are working hard to position the country as the "Bread Basket" of West Africa through her accelerated modernization and commercialization of agriculture, with women empowerment and re-orientation from subsistence production to market-oriented production. ... At the centre of the strategy is the empowerment of small, medium and large scale farmers ... to enable them[to] acquire and use appropriate modern technologies *to make farming in Ghana a business rather than a way of life.*[3]

The term "farming as a way of life" that the former minister of food and agriculture invoked refers to the practices of traditional[4] farmers in northern Ghana, which are often framed by development planners as reflecting isolation, insufficient production, a lack of knowledge, and a generalized "backwardness." With access to "right knowledge" (Grow Africa 2013: 12); new forms of agricultural extension; "sensitization"[5] seminars about land, seed, technology, and property; and market linkages, farmers can help to realize northern Ghana's food production potential. Farming as a business contrasts with farming as a way of life in its reliance on techno-scientific knowledge as well as its focus on profits and exports, rather than self-provisioning and traditional forms of community resilience. It also entails more formulaic farming (e.g., the use of prescribed inputs), rather than improvisation and experimentation (Richards 1993).

Farming as a way of life is also the commonplace practice of rural and peri-urban families producing their own food to feed their families. For example, plots adjacent to family compounds may grow grains (maize, millet, sorghum) alongside leguminous crops (cowpea, groundnut, Bambara beans), with various nutrient-rich greens for use in stews such as *bito*, *bra zheera*, and *ayoyo* grown on the perimeter of the plot. In the Upper East Region intercropping is still quite common (given its benefits in terms of the timing of harvests, weed and pest management, and protection from crops scalding), whereas the growth in the use of herbicides in the Northern Region has made intercropping incompatible where herbicides are utilized.[6] The use of implements such as the hoe is common, and some farmers also plow the land using oxen. Farming as a way of life does not preclude the sale of surplus to local markets—indeed that has a long history and continues to be practiced throughout northern Ghana. Seeds are saved and exchanged with other farmers, and sowing and other farming activities are timed in accordance to rain patterns. Trees of shea, *baba* (baobab), and *dawadawa* (locust bean, a "local Maggi") dot the landscape and are prized for their rich oils and flavor-enhancing properties. Farming is one aspect of a diversified livelihood strategy and may not be the primary occupation.

Another way of understanding what is meant by farming as a way of life is through an examination of the "non-market system," that is, "the system of exchange that exists in the local community" to build resilience that includes the exchange of seeds, food, and labor.[7] Informal systems of seed exchange are ways of diversifying the range of crops planted and forms of local innovation, frequently discounted in "Green Revolution" agricultural programs (Scoones and Thompson 2011: 16; Shiva 1991, 2001). Multiple informants in northern Ghana told me of the custom of sharing harvests with farmers who experienced harvest loss, delivering food discretely after sundown—a non-market form of food security. The practice of labor pooling is another important feature of the non-market system and is still an essential part of meeting labor shortfalls within certain families. The preparation of yam mounds in the Northern Region is one such labor-sharing practice. Groups of people from the community will focus their efforts on one family farm, rotating to another family farm the next day, followed by a few days' rest, although this practice is on the decline.[8] The sharing of traditional knowledge of crop cultivation and the use of plants is another important aspect of non-market systems in Ghana.[9]

From the Reproductive Power of Nature to the Productive Power of Inputs

Lakshman Yapa characterizes the Green Revolution as a fundamental shift in how food is produced:

> The Green Revolution created a technology which required poor farmers to buy inputs; it ignored other appropriate technologies of food production such as rain-fed farming, multiple cropping, growing of legumes, and so on. Its productivist logic marginalized political economists' concerns for people's access to land and productive resources. It devalued the "reproductive power" of nature by substituting the "productive power" of industrial inputs. ... By marginalizing traditional knowledge it robbed the culture of poor people of its power/agency to address problems of everyday life. (Yapa 1996a: 82)

His critique raises pertinent concerns regarding the promotion of farming as a business, as such a shift generates new wants (e.g., pesticides, herbicides, fertilizers, improved seed, tractors) and diminishes the value of existing resources and ecological services. That is, the shift from farming as a way of life to a business is contingent upon a move away from the reproductive power of nature to the productive power of industrial inputs (Yapa 1996a; Shiva 1991). This shift can be seen in the promotion of the following changes in agrarian practices in Ghana:[10] from the application of manure and cover cropping to the use of purchased chemical fertilizer; from tightly intercropped spaces that minimize the growth of non-beneficial plants to herbicides and pesticides that eradicate "weeds" and "pests"; from the saving and sharing of seed to the purchase of patented or hybrid seed each season; from the reliance on rain to the reliance on irrigation; from farming tied to seasons to agriculture that transcends seasonality; from farming activities powered by bodies (human and other animal) to machine power; from pastoralism and the integration of animals into farming to concentrated animal feeding operations; and from working with nature to working to control nature. Allow me to elaborate.

As discussed in Chapter 1, hybrid and genetically modified crop adoption create conditions, both biological and proprietary, respectively, that encourage the purchase of seeds each season. The purchase of improved seeds typically comes as part of a "package" with accompanying training and requirements of "seed-water-fertilizer" (Berry 1993: 199). As the interlude, "On 'Mixing,'" describes, seed saving and seed-exchanging practices continue to exist but are marginalized as more farmers embrace farming as a business. Farmers that show a willingness to try new technologies are recruited as "model" farmers, while others are integrated into agricultural value chain schemes through outgrower contracts that guarantee better market access.

With this transformation in political economy and technology, the landscape of the savannah shifts visually from "messy," intercropped spaces of permaculture to sole stands that promote monoculture and higher levels of production. Trees are reduced (or eliminated) as plots are cleared; they are considered obstacles to production that can be compensated for through monetary payment, rather than a community resource.[11] Attempts at continued intercropping are aborted when agro-chemicals used for maize production harm cowpeas that once occupied shared space. Traditional labor-pooling practices that functioned well when cultivation tasks were more flexible are complicated by the demanding timing constraints of hybrid varietals; this inflexibility makes labor shortages more acute where hybrid adoption is wider-spread (Berry 1993). As such, access to mechanization becomes more attractive. Irrigation becomes more important (and empty government promises of irrigation improvement more intolerable) with new varietals that have specific water-intake requirements.[12] Climate change makes reliance on rain-fed agriculture acutely challenging due to alterations in the distribution of rainfall.[13] Seasonality does less to circumscribe farming activities, as irrigation makes dry season farming possible and "market women" in the south of Ghana provide steady market demand.[14] Conflicts with pastoralists increase as livestock's damage to cash crops heightens the stakes of crop damage and more farmers resort to tying livestock to minimize this damage—and in the south of Ghana, concentrated animal feeding operations expand.

This shift from the reproductive power of nature to the productive power of industrial inputs denotes a move away from working with nature toward working to control it. Rather than rely on rain, irrigate. Rather than collect and save seed of crops that have performed well in the field, purchase seed that has been certified, has high market value, and is the product of laboratory research. Rather than allow animals to roam and fertilize fields as well as use them for labor, compensate the poor soil quality with fertilizer and mechanize. This suggests that the socio-ecological ramifications of the transition from farming as a way of life to a business may entail a loss of biodiversity in the seed system and in cropping patterns, increased dependence on fertilizers to support soil fertility,[15] and a decline in the care for the forest, seed, and labor commons. As such, the authority and value given to traditional agroecological knowledge—a critical element of community resilience—may decline, as farming professionalizes. It is to this dynamic that I now turn my attention.

The Professionalization of Farming

The professionalization of farming is central to the idea of farming as a business. Escobar (1995: 45) defines "professionalization" as

> the process that brings the Third World into the politics of expert knowledge and Western science in general. This is accomplished through a set of techniques, strategies, and disciplinary practices that organize the generation, validation, and diffusion of development knowledge, including the academic disciplines, methods of research and teaching, criteria of expertise, and manifold professional practices.

Building off of this concept, the professionalization of farming generates specific notions of what farming should entail, how it should be organized, from which knowledge bases it should be derived, and who should farm. (I return to the last point in my analysis of the logic of agricultural exit.) The professionalization of farming is also a way by which farming is taken out of the cultural and political domains and recast as a technical issue that requires new forms of knowledge and management to address (Escobar 1995: 45; Li 2007: 7–10; Mitchell 2002). Foreign and domestic development actors in Ghana, in their critique of farming as a way of life, situate themselves as having the unique knowledge, skill, and management necessary to transform agriculture in Ghana into a highly productive and lucrative industrial sector—that is, for farming to be "professionalized." This is an example of what Li (2007: 7) refers to as "rendering technical" where attention to the politics of agriculture (who has the land, capital, and power) is skirted by "focus[ing] more on the capacities of the poor than on the practices through which one social group impoverishes another." Questions of power and policy (particularly trade policy, as discussed in Chapter 1) are overlooked as development assistance and "capacity building" is welcomed to manage the technical problems of underproduction and farmer's lack of understanding of "good agricultural practices."[16]

The professionalization of farming is intended to foster growth in commercial agriculture in northern Ghana. The ways in which this is supposed to take hold is through increased linkages of farmers to regional and global markets, improved dissemination of knowledge about good agricultural practices, access to agricultural technology (improved seed, inputs, mechanization, and irrigation) to facilitate increased productivity, and better access to credit to support the "scaling up" of agriculture. Initiatives in northern Ghana that embody the professionalization of farming include: USAID-funded Agricultural Development and Value Chain Enhancement project (ADVANCE), Financing Ghanaian Agriculture Project (FinGAP), and the Agriculture Technology Transfer (ATT) Project; the World Bank–funded Ghana Commercial Agriculture Project; Masara N'Arziki; and the government of Ghana's Planting for Food and Jobs and Rearing for Food and Jobs programs. The farming practices of lead farmers enrolled in these agricultural development programs encourage emulation (Stone 2007; Henrich 2001) through various communication strategies: demonstration farms, peer influence, films, and radio jingles. Professionalization of farming entails an expanded role for the private sector, not only in the provision of inputs but also through the privatization of agricultural extension. As farmers in Ghana become more integrated into agricultural value chains, the private sector increasingly gains prominence as a locus of agricultural knowledge dissemination.

The integration of farmers into value chains via outgrower contracts is one way that farming can be professionalized. A value chain is based on the idea of value addition and specialization in commerce. The integration of smallholder farmers into global and regional value chains is perceived by development planners as inherently positive, as the local markets (wherein they are acknowledged to exist) are believed to be inadequate, and it is this inadequacy that is believed to be at the root of the poverty that persists within African agrarian economies. The conceptualization of global value chains envisions a series of integrated and interconnected productive and consumptive actions that link smallholder farmers to "nucleus" or "progressive" farmers that are then linked to input and seed suppliers, to processors and aggregators, to supermarkets and other retailers, and, finally, to consumers. A 2010 World Bank report, *Building Competitiveness in Africa's Agriculture*, specifies that outgrower schemes are generally understood as

> schemes where agribusiness has considerable control over the smallholder production process, providing a large number of services, such as input credits, tillage, spraying, and harvesting. The smallholder provides land and labor in return for this comprehensive extension/input package. (Webber and Labaste 2010: 133)

These outgrower schemes are generally linked to global value chains (Gibbon and Ponte 2005), which facilitate smallholder farmer's access to financial and informational services. For example, smallholder farmers may produce cash crops as outgrowers under contract to nucleus farmers or aggregators, who then sell their products to much larger processors that add value to the raw materials and sell to suppliers and wholesalers. Typically, nucleus farmers or aggregators supply inputs and access to tractors to smallholder farmer "outgrowers," who then pay back this "interest-free loan" with in-kind agricultural produce.[17] These linkages are intended to assist in the identification of needed "improvements in supporting services and the business environment" (Webber and Labaste 2010: 1).

A final, and increasingly important, aspect of the professionalization of farming in Ghana is found in higher education. In 2007 the West African Centre for Crop Improvement (WACCI) at the University of Ghana, Legon, was founded under the leadership of Dr. Eric Danquah with $11.2 million in funding from the Alliance for a Green Revolution in Africa. This is a sister institution to the African Centre for Crop Improvement at the University of KwaZulu-Natal in South Africa. Instructional support provided by Cornell University and seed money from AGRA helped to establish a PhD program in Plant Breeding, and as of August 2018, they had enrolled 114 students and graduated 66.[18] WACCI markets itself as "Training the Next Generation of Plant Breeders" and "Developing Improved Crop Varieties to Feed Africa" and has secured the World Bank's support to make it a Center of Excellence. Since WACCI's establishment, two master's degrees have been developed at the University of Ghana, Legon: an MS in Seed Science and Technology supported by Iowa State University and an MS in Genetics and Plant Breeding supported by the Soybean Innovation Lab (SIL) out of the University of Illinois, Urbana-Champaign.[19] According to an informant working with the SIL, the MS in Seed Science and Technology has received the enthusiastic support of the World Bank, "who loves to fund graduate training."[20] These programs are costly and exclusive, requiring years of work experience in agricultural research in Africa. To mitigate the costs, the Alliance for a Green Revolution in Africa and the Syngenta Foundation offer plant breeding scholarships; the ability to secure external funding is a criterion considered for acceptance (WACCI n.d.a).[21]

An examination of these programs reveals the ways in which universities and biotechnology research centers in the United States may shape how agricultural sciences are taught in Ghana, as well as other African countries. Universities that include Cornell University; the University of Illinois, Urbana-Champaign; Iowa State University; University of California, Davis; and Purdue University, among others, shape higher education curriculum in Ghana through their expertise and emphases on plant genetics, biotechnology, and molecular plant breeding, as well as their capacity to reach students through internship opportunities and e-learning workshops. As part of the MS degree in Genetics and Plant Breeding, select students have the opportunity to intern in "advanced laboratories in the USA for two months experiential learning and skills development in Plant Breeding" prior to undertaking their thesis research (WACCI n.d.b). Iowa State University partnered with the Kwame Nkrumah University of Science and Technology to launch a week-long "E-learning Workshop for Plant Breeding Educators" in March 2019 that drew Ghanaian participants from the University of Cape Coast, the University for Development Studies, and the University of Education, Winneba, and was supported by AGRA with funding from the Bill & Melinda Gates Foundation (Plant Breeding E-Learning in Africa n.d). The MS degree in Seed Science and Technology at WACCI provides "students with current seed science and technology instruction along with essential modules in business management in a rigorous, integrated curriculum to develop a network of broadly trained individuals that can work effectively to improve access to high quality seeds" (WACCI n.d.b). A key component of this network is entrepreneurs that can work alongside this new crop of experts, which I turn to next.

The Promotion of Entrepreneurship and Agricultural Risk-Taking

The promotion of entrepreneurship is complementary to the professionalization of farming. An entrepreneurial farmer is one that sees farming as a business and is able to envision scaling up production to transition Ghana to become the "bread basket of West Africa." The entrepreneur is concerned with profits, produces cash crops for export, and is able to meet certification requirements. They likely come from a family that owns land, perhaps from a chiefly family, and would be considered a "big man," a person of status, for their business know-how.[22] Rather than rely on communal systems of reciprocal labor exchange during the harvest season, they hire paid laborers to work on their farm. They have received training in the use of inputs, utilize mechanization, and know good agricultural practices. They are a part of a value chain and know how to access capital to scale up their business. They are willing to take risks. They keep up with new technologies and are eager to use them. They have the large commercial farm that is connected to smaller outgrowers that sell their product in kind for the use of their tractor. For an example of this promotion of entrepreneurship, allow me to share a story.

***

We had walked in late to the Mango Value Chain workshop, but it didn't seem to bother Latif, who had just been enskinned[23] as a chief. My husband and I had met him through a mutual friend, and Latif was representing his brother's company, one of the largest agro-chemical dealers in Tamale. It was an uncomfortable affiliation, as Latif filled out the sign-up sheet with our names as agro-chemical affiliates, and one that I was fortunately able to distance myself from when we joined the forty participants—I was one of four women in the room—who had gathered to talk about how to develop and strengthen the mango value chain in northern Ghana. The workshop was put together by the Market-Oriented Agricultural Programme (MOAP), a German-Ghanaian PPP. It was quite obvious that there was German money involved—everyone was given a nametag, a spiral notebook, a pen, and a plastic folder. Bottles of water were at each table in the air-conditioned room, with breaks of fish pies and sodas and a full spread of food at lunch.

After the first coffee break, Nana Ampofo spoke about the experience of mango production in Brong-Ahafo, a region directly south of the Northern Region, which had well-established mango plantations and is the center of mango production in Ghana. In a talk entitled "Lessons from Brong-Ahafo," Nana spoke of the need to "target consumers first" and posed as a central question, "what does the consumer want?" The identification of the "consumer" became clear with his follow-up question, "how can mangoes be exported?" Whereas he later addressed local consumer demand, his clear focus was the production for export markets in locations like the European Union. Growers, Nana explained, need to acquire more knowledge of mango production and should look to input providers for fertilizers, agro-chemicals, and seedlings. Service providers such as banks also played a role in strengthening value chains. Nana Ampofo spoke of issues like postharvest loss, certification, the necessity to establish standardized good agricultural practices in Ghana, the damage posed by fruit flies, and the need to use caution when using insecticidal sprays. He then emphasized the importance of training—farmers need to be taught how to spray and how to handle pesticides. He also emphasized that the lack of access to information was a key impediment to success: going into "serious mango business" requires information.

During the question-and-answer session that followed, Nana Ampofo stated, "We have a motto that says 'mango is a serious business.' ... If you don't want to get into a serious business, then don't get into mango." One of the facilitators of the event, Al-Hassan, found this articulation of mango farming as a "serious business" so compelling that he led a call and response whereby he would shout "mango" and we participants would respond "serious business!" Al-Hassan explained that this refrain was important because it expresses the idea that producing mangos for export (that meet international standards) is a different enterprise than meeting local demand. The workshop included value chain role-playing activities to communicate some of the common challenges that different actors along the mango value chain experience, as well as opportunities to discuss possible solutions. Throughout the nine-hour workshop, a tension arose between producers who thought largely in terms of local markets and service providers in the business of export promotion like EDAIF (Export Development and Agricultural Investment Fund). In these hours of exchange, participants frequently talked past one another, and producers expressed the kind of frustration of persons that have expressed their concerns, but have not been heard. As one producer explained—much to the chagrin of organizers— farmers must produce for themselves *first* and then for international markets. Sentiments such as these were not the intended outcome of the Market-Oriented Agricultural Programme.

***

Commercial agriculture programs, such as MOAP's support of the "mango value chain," communicate that getting "serious" about farming requires information, networking, entrepreneurship, and the ability to link into value chains. Participation in these production schemes mitigates risk through access to (1) secure markets to sell products, (2) credit, and (3) agricultural technologies. However, the serious farmer is also an entrepreneur willing to take risks—adopt new agricultural technologies, scale up through accessing credit—for the potential of great gains to productivity and profits. Interventions such as ADVANCE, ATT, and Planting for Food and Jobs encourage risk-taking by reducing some of the risks of adoption of technology:

We have a small grant component in which we diminish the risk of adoption of technology. Sometimes a farmer won't go into something just because they've never used it before, like threshers. Threshers are a little bit expensive, extremely profitable, but still they don't have $8,000 dollars, $6,000 dollars to invest in them. So, we give 70% of the value of the equipment, and they pay the other 30%.[24]

While the risk-sharing design of this project may encourage technology adoption, it is likely the case that it would be the larger farms that would be the beneficiaries because they can better handle the adoption risks (Collier and Dercon 2014: 94–5). That is to say that larger farm enterprises "can afford to use some of its plots for trial and error—and then adopt soon afterward the successful innovations" (Collier and Dercon 2014: 95).

By contrast, traditional farmers typically operate by a "safety-first" principle whose goal is a secure subsistence (Scott 1976). This goal "is expressed in a wide array of choices in the production process; a preference for crops that can be eaten over crops that must be sold, an inclination to employ several seed varieties in order to spread risks, a preference for varieties with stable if modest yields" (Scott 1976: 23). Scott (1976: 25) contends that the self-provisioning farmer is risk averse and hesitant to adopt new technologies because "he works close enough to the margin that he has a great deal to lose by miscalculating." By this logic, stable yields are more important than yield increases; new technologies that promise yield boosts may be regarded skeptically without the environmental learning[25] that can demonstrate the new technology's stability over time. Subsistence farmers further reduce their risks and improve food security through the labor-sharing practices and systems of exchange of seeds and food described earlier.

This safety-first principle is reflected not only in the choice of which crops to grow but also in the prioritization of efforts. As one of the producers explained in the Mango Value Chain workshop, producing for the local market and community takes precedence over production for international markets. Farmers' associations have been frustrated by the frequency with which smallholder and tenant farmers divert inputs like fertilizers and pesticides (intended for the maize that has been purchased through the association) to tend to the crops produced for family consumption.[26] Landowners in the Brong-Ahafo region have tried to discourage their sharecroppers from intercropping—a commonplace subsistence practice to get through to harvest—arguing that the pruning damages the cash crop, and it is not part of tenants' rights (Nyantakyi-Frimpong and Bezner Kerr 2017). Such observations are consistent with Scott's account that "the villager only attends to his 'selling rice' field *after* his subsistence field tasks are complete" (Scott 1976: 23; emphasis in the original). Following the devastation of tomato crops brought on by roundworms called nematodes (resulting from poor crop rotation), one of the chiefs I spoke to in the Upper East, a farmer, explained that the rush to plant tomatoes was about "quick money" and criticized these farmers' failure to diversify their farming. "This is not good for food security. ... They should be planting crops that you can store ... maize, rice, groundnut ... not perishable crops."[27] The safety-first principle is being challenged with the expansion of commercial agriculture, but, as these complaints of the failure to heed traditional wisdom reveal (as well as that of many other conversations I had in the Upper East and Northern Regions), feeding the family first is still of preeminent concern to many people in northern Ghana.

Events like the Mango Value Chain workshop reorient farmers' attention to export markets and are premised upon the role of smallholder farmers as agents of development. By modeling good agricultural practices and being an "early technology adopter," serious farmers create greater demand for commercial seed through visual evidence of the value of these agricultural technologies by the success of their farms. These serious farmers are the "others, women and men, [that] have pursued different options to escape poverty" and rejected a life of subsistence (World Bank 2007: 1). As one informant working with ADVANCE told me, "Farming needs to be seen as a 'big man's job.' "[28] Another farmer, whose father owns one of the larger farms in the Northern Region, told me that farming is often associated with being "poor, illiterate" and not considered a profitable enterprise, contributing to the loss of young people in agriculture.[29] The thrust of these commercial agriculture programs in Ghana is to change peoples' perceptions about the nature of farming: farming can be a business, and a profitable business at that. Training events like the Mango Value Chain workshop, which are taking place all over Africa, are more than just opportunities to train farmers in the meaning of good agricultural practices. They are attempts to create new identities. When the adoption of new agricultural practices is framed as part of "progress," this adoption produces corollary subjects that are "modern." Likewise, the rejection of new agricultural technologies can render those that are averse to these technologies as "backward" and provide justification for land dispossession, as I discuss in the section on the logic of "agricultural exit." Next, I talk about how efforts to promote the commercial seed sector in northern Ghana attempt to reach the last mile user that will adopt new technologies and play a part in this agricultural transformation.

III. The Last Mile User and the Promotion of a Commercial Seed Sector in Northern Ghana

You take it to the last mile user. That farmer in that community. You go there and you find everything in that community except [certified] seed.[30] ... Make it available, just as we are making the crop protection product available in their community. Just as we are taking fertilizer to their community. Take seed there and let's see that farmers will use it. Advertise it, and let's see if the farmers will use it or not.

—Project Manager of the USAID ATT Project, July 27, 2018

It is tempting to conflate the last mile user with the farmer in interpreting the meaning of this statement. However, doing so obscures how a different set of relations— social, economic, and ecological—are mobilized whereby a person is reduced to a "user" (that follows a prescription of inputs), rather than a "farmer." The last mile user imaginary is also a vision of customers that have not been met, agricultural frontiers of capital untapped, technological solutions to problems yearning for an answer.[31] In conceptualizing the delivery of agricultural technologies to the last mile user, local, traditional seed is there, but unseen by development planners. (If it were not there, then there would be no farming in these communities defined by subsistence agriculture.) Here, traditional seed is not of interest, not qualifying as "seed" in this official's perspective, and not of obvious value and rationale.

John Awedoba, who is a "nucleus farmer" in Bolgatanga for Feed the Future (and one of the "mixers" portrayed in the following interlude), makes a key distinction between this introduction of certified seed and traditional crops. He explained that while the sale of these cash crops brings in income, maize and rice are "high feeder crops," meaning that they require high levels of applications of fertilizers. By contrast, traditional crops such as millet, sorghum, cowpea, or Bambara beans are "low feeder crops that need less inputs." He added that the soils for the high feeder crops require increased application of fertilizer over time, which is both costly and continuous for successful crop production.

The needs of these "high feeder crops" are echoed in this elaboration of the ATT Project:

> We need to convince farmers of their need to use certified seed. … So we are not only looking at seed as a technology, but we also look at the integrators of activity management, so to move around with the seed, you need the right fertilizer recommendation and other stuff for the seed that you're promoting.[32]

The core objective of the ATT Project, funded by USAID and implemented by the International Fertilizer Development Center (IFDC), is to "increase availability and use of agricultural technologies to increase and sustain productivity in northern Ghana" (Feed the Future n.d.: 2). Interventions combine a focus on demonstrations, training, audio and video extension, technology fairs and promotional materials, mobile vans, and business development services to promote certified seed, integrated soil fertility management, and labor-saving agricultural technologies.[33] In its five years of operation, the ATT Project reached 173,441 smallholder farmers through forty-three public–private partnerships and US $5.7 million in grants to beneficiaries (Feed the Future n.d.). It was assessed by the Center for Strategic and International Studies Global Food Security Project as "catalyzing big and important change in the seed sector, from development to distribution" (Cooke and Flowers 2018).

Twenty thousand "starter packs" of seed, fertilizer, and inoculant have been distributed for free on the basis of a "try it and see" model coupled with audio and visual extension and "seed vans" that enable local agribusiness to reach the last mile user.[34] Both the starter packs and the seed vans are paid for by matching grants: the private sector (e.g., in the Northern Region, Wumpini Agro-Chemical and Heritage Seeds) provides 70 percent of the funding for the starter packs while the ATT Project provides 30 percent; for the seed vans, this is reversed, with ATT paying 70 percent and the private sector beneficiary 30 percent.[35] Seed distribution vans were highlighted as a unique innovation:

> The seed producers have already made their own jingles, and they go to communities [in their seed vans], in their local language they are speaking to them, talking to them about the importance of certified seed. And every community they go to, the community folks are running around that van and coming to purchase seeds.[36]

Through partnerships,[37] the ATT Project also brought films to villages via projector backpacks that screened local language films featuring model farmers discussing their success with new agricultural technologies that include certified seed. The "digital classroom" strategy is part of a "new extension model" that addresses the "inadequate reach" of government extension by turning increasingly toward USAID-supported extension, whose content is developed in partnership with the private sector and additionally supported by philanthrocapitalist funding.[38] This work is envisioned as a means to a commercial seed sector in northern Ghana, although some development experts recognized that it is "non-trivial to move away from seed saving."[39] It is also critically a part of this transition from farming as a way of life to farming as a business.

Whereas the ATT Project selects input providers and seed growers to receive labor-saving equipment such as seed processing machinery and seed distribution vans to reach the last mile user (an example of philanthropy for an emergent private sector), these local private-sector actors are also taking a risk by sharing in the cost through these matching grants. ADVANCE, another USAID-sponsored program to support the expansion of commercial agriculture in northern Ghana, has a similar approach, as one official explained,

> We don't give anything to the smallholder farmers. We put them in touch with the private sector. The demos that we establish, we don't buy a single input. The private sector comes and sponsors it, that's the way they market their products.[40]

While the audio and visual extension provided by the ATT Project is shaped by the research funded by IFDC and USAID, it is utilizing materials provided by the private sector in the form of these starter packs. My research reveals a shift in the nature of extension, not only away from the public sector (even as it is reliant upon it to conduct trainings, as discussed in Chapter 2) but increasingly toward the private sector as a locus of knowledge dissemination. Conflicts of interest—given that there is a perverse incentive for agro-chemical and seed distributors to encourage increased consumption of their products—of private sector extension were noted in these interviews, but there was no real address of the issue.

Further upstream is the role of development experts that generate the products for this new form of extension, such as the University of Illinois SIL's "soybean success kit" featuring pictorial extension used to train development practitioners that include ATT and ADVANCE. Principal investigator of SIL, Peter Goldsmith, brings his expertise analyzing (and promoting) tropical soybean industrial development in Brazil to sub-Saharan Africa, with Ghana being at the "core" of the "influence zone" for examining the "appropriateness of soybean in sub-Saharan Africa" since 2013.[41] Training such as the MS Degree in Seed Science and Technology at WACCI "leaves the entrepreneurial class to do the work of seed development ... managing companies, research stations, seed production."[42] An expert involved in the curriculum development saw a lack of an "entrepreneurial class" in Ghana that can manage seed development and viewed the SIL as an opportunity to develop it. "FDI [in Ghana] doesn't occur because of this lack of managers."[43] As previously discussed, a central strategy of attracting foreign direct investment has been to create an "enabling environment" for investment, and this informant stressed the absence of managers as a reason for Ghana's lack of investment. Furthermore, the development of an entrepreneurial-managerial class works hand in hand with privatized regulation to attract global investment. AGRA and ATT together have promoted the patenting of seed and intellectual property rights and are "gaining ground to release some of those regulatory policies to the private sector."[44] AGRA's policy advisory support helped to shape the 2010 Ghana Plants and Fertilizer Act (as discussed in Chapter 3), which created a provision for private seed companies to produce foundation seed—previously the exclusive domain of the public sector[45] (Kuhlmann and Zhou 2016: 6; AGRA-SSTP n.d.).

New Opportunities, New Vulnerabilities

ADVANCE links farmers to the private sector—actors that include Yara, RMG, AGRICARE Ltd.—as well as to information such as weather forecasting (Ignitia) and market information (Esoko Market Information).[46] In a similar vein, ATT's end-of-project evaluation showed that some of its greatest successes were in the number of farmers accessing marketing and technology information on seed, integrated soil fertility management, and agricultural practices through ICT (information and communications technology) mechanisms as well as in the "value of incremental sales of targeted ATT commodities (seed, fertilizers, and other soil amendments)" attributed to Feed the Future implementation (Feed the Future n.d.: 6–7). Sales of targeted commodities have increased substantially, and to an even greater extent, so has the use of text messaging, digital classrooms, and other ICT mechanisms to provide information and this new form of agricultural extension (ibid.). This means that there has been a significant increase in the uptake of certified seed—for the time being[47]—and in the reliance of farmers on weather forecasting and market information to make agricultural decisions.

At the 2019 Planting for Food and Jobs campaign launch, the minister of food and agriculture attributed the bumper harvests that enabled agricultural exports to neighboring countries to the success of the campaign's subsidized seed as well as crop- and agroecological-zone-specific fertilizer. *Planting for Food and Jobs: Strategic Plan for Implementation (2017–2020)* states that the plan "will motivate the farmers to adopt certified seeds and fertilizers through a private sector led marketing framework, by raising the incentives and complimentary service provisions on the usage of inputs, good agronomic practices, marketing of outputs over an E-Agriculture platform" (Ministry of Food and Agriculture Ghana 2017: 3). Proponents of this government program argue that Planting for Food and Jobs is different than previous failed efforts to develop commercial agriculture in northern Ghana. It is "private sector-led," it is supported through PPPs such as AGRA's capacity-building support of the Soil Research Institute, it facilitates agricultural extension by way of a digital platform, and it allows farmers to split payments for inputs (pre- and postharvest). Private sector-led interventions were considered beneficial not only because of a fierce belief in the innovation and problem-solving capacity of the private sector but also because of the acknowledgment that project funding often ends before there is any effect. That is to say that while the projects have finite start and end dates, a private sector with a substantial base of customer support can continue the objectives of these projects into the future.

Although these projects share in the risks, there are ways in which these projects may encourage risk-taking behavior that may not pay off. The ATT Project notes one of its successes as facilitating US $2.23 million in loans from financial institutions. However, increased access to credit by lead farmers and small businesses is not without risk. An ADVANCE official characterized smallholder farmers' participation in ADVANCE's outgrower contract program by high rates of default, as much as 50 percent some years.[48] I frequently heard accounts about the risks that these farmers would need to take on as outgrowers, from varieties of mango ill-suited for the northern Ghanaian climate but desirable for export, to undependable rainfall patterns, to the possibility of large quantities of crops being rejected when not meeting the proper specifications for stringent export markets. These defaults not only reflect an outgrower's struggle to secure his or her livelihood but also directly impact the nucleus farmer as he or she awaits in-kind payment that may adversely impact his or her ability to sell to grain aggregators.[49]

Take as another example the Ghana Commercial Agriculture Project. GCAP is structured in such a way that the "commercial farmer"[50] is selected for access to large tracts of land and receives a matching grant "in the form of land development."[51] The commercial farmer is responsible for the inputs and services (land preparation, seed, fertilizer, and agro-chemicals) used for commercial outgrower schemes that outgrower farmers pay back generally with in-kind grain. USAID's FinGAP facilitates this access to finance, but similar to ADVANCE, there have been defaults by outgrowers when yields underperform due to erratic rainfall, in ways that "you cannot fault them."[52] The commercial farmers are nevertheless responsible for this repayment of debts, regardless of drought or devastating storms.

In northern Ghana, GCAP's "land facilitation mechanism" assigns suitable "underutilized" land identified through expert assessment and a competitive application process to entrepreneurial farmers nationwide.[53] GCAP is also a World Bank–funded project, and a portion of the funds secured by the Ghanaian government in 2015 were to rehabilitate four dams, including the Vea Dam.[54] When I visited Nyariga, a community reliant upon the Vea Dam for irrigation, in the summer of 2018, the people I spoke with complained that no government work had been done on the dam—other than the placement of GCAP Vea Dam Rehabilitation signboards that stood rusted and sun-faded adjacent to where we were talking—and that even with attempting to remove silt from the dam by hand,[55] the water failed to reach their community. One lead farmer, Idrissu Alkobiri, told me he was contracted to grow an AGRA rice varietal through GCAP and stated that he initially liked growing this rice because it was popular for its "nice" taste, fragrance and shorter preparation time. Growing this rice allowed him "to make a profit" when the rains were good, but due to recent erratic rainfall, the yields on his farm and that of his outgrowers were so poor that he fell into debt. He complained that he still had to pay back his debt, "even when they come to see that the crop has failed" and that the risk fell entirely on them: "the contractor doesn't bear the risk."[56]

IV. Experts, Entrepreneurs, and the Logic of Agricultural Exit

Interventions like the ATT Project, GCAP, and Planting for Food and Jobs are intended to connect farmers to the tools necessary (both material and informational) for increasing aggregate yields. They aim to transform an agricultural landscape rife with subsistence farming into a "bread basket" of high agricultural productivity through "pro-poor private sector led value chain development" (Republic of Ghana 2014: 4). Many of these types of interventions in rural Africa identify an aging population of farmers and a declining interest in farming among young people. In order to address these dynamics, agricultural interventions need to not only change the existing orientations of farmers from producing for households to producing for export-oriented value chains but also entice young people into farming. Presenting commercial agriculture as a lucrative enterprise that can rely on mechanization, rather than difficult manual labor, is intended to attract young people into food production. PPPs have played an influential role in altering such orientations through a discourse of entrepreneurship.

The African Accelerated Agribusiness and Agro-Industries Development Initiative, otherwise known as 3ADI+, is a collaboration between the United Nations Industrial Development Organization (UNIDO), the UN Food and Agricultural Organization (FAO), and the International Fund for Agricultural Development (IFAD). As stated on the UNIDO website, through assisting in the processing of agricultural commodities (one of the links in the value chain), 3ADI+ "can help them make the leap from subsistence agriculture to a thriving business that generates income." The objective is to "transform the rural world to turn it into an attractive career proposition to the eyes of the youth" (ibid.). In this way, farming becomes "no longer a subsistence occupation carried out from generation to generation as a matter of tradition: it is a complex business with its technological, scientific, human resource, marketing, and accounting demands" (ibid.).

What kinds of people do the proponents of agricultural modernization want in agriculture? Educated youth, trained technocrats that can use modern farm inputs, serious farmers that view farming as a business. In such efforts to modernize agriculture, "unproductive," subsistence farmers are identified as those who should "transition" out of agriculture. Scaling up is a concern of these agricultural development programs due to the expectation that less young people will choose rural occupations like farming and due to an aging population of farmers. Farmers' associations and the placement of graduates of agriculture programs in the field are ways to train young people to become serious farmers. A collaboration between Premium Foods, a Ghanaian processing company; the African Union; and the Kwame Nkrumah University of Science and Technology (KNUST) plans to take graduates of the KNUST agriculture and agribusiness program, allot them 100 acres, and connect them to "champion" farmers who will be given 5 acre parcels of land. The hope is that the university students' technical knowledge will lead to change in practices on the farm.[57]

How do these development interventions identify serious farmers to play the part as lead farmers in these value chain schemes? While informants consistently communicated that they choose farmers who are "respected in their communities" and "opinion leaders" that had land, I observed that there was a common pattern of some of the same farmers being chosen to be a "model farm" for various development interventions.[58] When interviewing a project manager of the ATT Project in Tamale about how they choose which farmers to work with to lead field demonstrations that utilize new agricultural technologies, he stated, "We've been in this environment for quite a long time, and we know the key guys around."[59] This statement reveals two things: first, that these development interventions have been around for "quite a long time," suggesting that the work being done is not all that novel, and second, that they likely work with the same "key guys around." This is the case with the farmer entrepreneur Al-Haji Zakaria, owner of Heritage Seeds in Tamale, who has been chosen to work with the Savannah Agricultural Research Institute on a nucleus-outgrower scheme for cowpea propagation.[60] His company has received "technical and financial support ... to produce certified seeds, create awareness among farmers on the new seeds and distribute the new seeds to farmers" from the Alliance for a Green Revolution in Africa and is also the recipient of an ATT seed van and equipment from USAID, described in the section on the last mile user (AGRA n.d). Far from suggesting that Zakaria is undeserving of such support,[61] the point is to highlight the ways in which development actors select certain farmers to model new agricultural practices that have already received support and recognition from other development actors. In a sense, development is targeted at those that have already been "vetted."

What, then, are the mechanisms by which this transformation from farming as a way of life to a business takes place? Training events, farmers' associations, access to credit, and the linkage of smallholder plots with larger nucleus farms are some of the mechanisms.[62] Training events like the Mango Value Chain workshop in Ghana are hosted by export-oriented service providers, input suppliers, or the Ministry of Food and Agriculture extension officers. Farmers' associations like Masara N'Arziki are programs that provide a package of inputs and a ready market for maize. On a billboard on the northern edge of Tamale along the Bolgatanga road, Masara N'Arziki advertises: "Masara Your Association for: better bargaining power; training and education; improved technology; guaranteed markets; sustainable access to credit; higher yields and higher income." Similar programs in Burkina Faso, Ghana's northern neighbor, have facilitated the expansion of genetically modified cotton.[63] Such arrangements encourage standardized agricultural practices and shift the focus of agricultural production to certain commodities. In a similar vein, the nonprofit international economic development program ACDI/VOCA, which has supported value chain development programs in Ghana, explains that their farming as a business curriculum

> creates profound change in a smallholder farmer's mentality and in her or his prospects for economic improvement. It is not merely a new way of thinking, but also a substantially novel way of operating that puts even a small farm on an enterprise plane and provides tools for proper management. (ACDI/VOCA n.d.)

With proper management and mentality, farmers can meet the expectations of export markets.

Private sector–led farm tours and the USDA's exchange programs have become other ways by which knowledge of commercial agriculture is shared with Ghanaian farmers. Tours of places like Grand Forks, North Dakota, led by groups like Praxis Strategy and AdFarm "focused on identifying and developing opportunities in the Ghanaian agricultural sector" (Pates 2011). Through touring commercial farms in the United States, participants are encouraged to envision parallel agricultural developments within their home country. Praxis Africa, one of the outcomes of the collaboration between Praxis Strategy and AdFarm, has developed a Farm Channel that disseminates information in Ghana on how to develop value chains to "promote farm productivity and profitability in Ghana" (Praxis Africa n.d.). The USDA Cochran fellowship is a three-week capacity-building exchange program that sends 50–75 Ghanaians to the United States, where they work with US universities and learn about food safety, feed formulation, and US agricultural cooperatives.[64] Such experiences enable participants to envision the various steps of commercialization and scaling up of agricultural production. Yet, conversations I have had over the years with Ghanaian farmers, agricultural research scientists, and biosafety regulators have indicated that the products of American agriculture left something to be desired.[65]

Agricultural Exit

A report published by the influential DC-based agricultural think tank International Food Policy Research Institute (IFPRI) articulates both the "key role" "smallholder farmers" have in meeting "future food demands" but also qualifies what type of smallholder farmers they have in mind:

> However, smallholders are not a homogeneous group that should be supported at all costs. *Whereas some smallholder farmers have the potential to undertake profitable commercial activities in the agricultural sector, others should be*

supported in exiting agriculture and seeking nonfarm employment opportunities.
For smallholder farmers with profit potential, their ability to be successful is
hampered by such challenges as climate change, price shocks, limited financing
options, and inadequate access to healthy and nutritious food. By overcoming
these challenges, smallholders can move from subsistence to commercially
oriented agricultural systems, increase their profits, and operate at an efficient
scale—thereby helping to do their part in feeding the world's hungry. (Fan,
Brzeska, Keyzer, and Halsema 2013: 16)[66]

What are the implications of this development approach for smallholder farmers
deemed not to have "profit potential"? And who decides that? The statement above
by these experts suggests that such farmers should be aided in transitioning out
of agriculture altogether. The World Bank development report *Agriculture for
Development* echoes a similar sentiment:

> Agricultural growth is especially important to improve well-being in geographic
> pockets of poverty with good agricultural potential. For regions without such
> potential, the transition out of agriculture and the provision of environmental
> services offer better prospects. But support to the agricultural component of the
> livelihoods of subsistence farmers will remain an imperative for many years.
> (World Bank 2007: 22)

Those farmers who "under-produce" are expected to shift out of farming into the
wage labor economy as they are not suitable for the project of "feeding the world."[67]

As Amanor (2009) points out, such decisions to sell land and enter the wage
economy are presented as an individual choice to alleviate poverty rather than
evidence of a lack of options. One informant working on the SADA initiative
commented that the local farmers' participation in proposed commercial
agriculture projects like sugarcane for biofuel will likely be negligible because of
the inability of such farmers to operate on such scale. Rather, "local farmers will
be turned into laborers" to work on those plantations.[68] Through this shift from
being in control of the farm to working as a laborer, local farmers may "forget
those skills" and may undermine the viability of farming as a way of life for future
generations.[69]

This logic of agricultural exit put forth by these development planners can be
thought of in Foucauldian terms as the neoliberal state's exercise of the power of
regularization. As part of the normalization of industrial agriculture, the outliers
of the state's population—in this case the "unproductive" small farmers of the
North—can be sacrificed.[70] This can help us understand why certain biotech and
hybrid seed proponents dismiss the critique that the cost of inputs to support the
use of improved seed is too high.[71] That is, if these farmers are just outliers in a
population that gets normalized around the ideal of farmer-as-entrepreneur, the
state is not necessarily going to intervene to bring these farmers in line with the
norm by making these inputs affordable for them. Rather, they will be outliers
on this curve of normalization that the state will "let die," or exit agriculture.

This exercise of the state power of regularization and the logic of agricultural exit would be a lot more worrisome if this outlier population were so dependent on the Ghanaian nation-state. Time spent in rural villages in northern Ghana reveal that many such communities rely little on state resources, but rather practice local forms of risk reduction, food security, and communal support.

V. Conclusion

Formal education, educational exchanges, and training normalize industrial agriculture and build trust in the ability of expert knowledge to address agricultural challenges. Western, industrialized agriculture is perceived as the model, and African agriculture is framed as deficient in this light. Rather than asking what can be learned from traditional agricultural practices, many Western-educated young people no longer view the village as a rich source of knowledge. By contrast, young people with access to education now return to the village as the "experts" to educate the elders. Furthermore, the patterns of who becomes the entrepreneur frequently replicate existing North–South divides in Ghana—it is the college-educated KNUST student from Kumasi that is the model farmer, the entrepreneurial farmer who wins a nation-wide competition to invest in land in the North as part of GCAP (but who may otherwise have little knowledge or connection to the agricultural community there), or the serious mango farmer from Brong-Ahafo. That is to say that the work of entrepreneurial farmers and their nucleus farms and demonstration fields can inspire farmers to become serious, emulate their behavior, and one day become a farmer like them. Yet, few farmers in these programs are likely to realize this ultimate goal. The more common prospect is to find themselves in the position of outgrowers with fairly significant constraints on their activities. If the last mile user approach takes hold, local agro-chemical and seed distributors will have increased local demand due to the high-input requirements of certified seed. Whether that demand will be met by local or global agribusiness remains to be seen.

The Ghana Commercial Agriculture Project recognizes that its program is not intended for subsistence farmers "farming under extremely fragile and disadvantaged circumstances" and points to USAID's Resiliency in Northern Ghana Project as one that addresses the needs of the most vulnerable farmers (GCAP n.d). Fair enough. But how do GCAP and other farming as a business programs address vulnerabilities among farmers that are a part of these agricultural schemes? How are the effects of climate change, which can decimate harvests through destructive rains, winds, and drought, being addressed[72] as farmers are encouraged to take on new risks of technological and financial innovation? What is the appropriate remedy for heightened economic precarity given the interactive effects of climate destruction and inadequate irrigation infrastructure as nucleus farmers are encouraged to take on debt? These stories of default among outgrowers, high interest rates (and their improper disclosure[73]), indebtedness, and climate change should urge caution about the allure of farmers' access to finance, central instruments that facilitate access to inputs and mechanization in nucleus-outgrower contracts.

Analysis of this period of agricultural transformation allows us to see within it struggles over knowledge, identity, and authority. How should farming be organized? What does it mean to be a successful farmer? Who should determine how farming should be practiced? Who possesses the knowledge best suited to support agrarian livelihoods in northern Ghana? Efforts to develop an entrepreneurial-managerial class in Ghana shift these identities (from farmer to user), locus of knowledge, and even responsibilities of governments as extension transforms. As the interlude reveals next, although there are interventions to render legible farming to support the expansion of the market economy, the multiple roles that people in rural northern Ghana perform, their food distribution and risk-reduction strategies, and their tendencies toward a pragmatic hybridization of farming practices suggest that efforts to standardize, professionalize, and integrate farming into global value chains will encounter obstacles. However, there is one significant trend that presents the possibility that a "new Green Revolution" may take hold in northern Ghana: the shift from elders being the bearers of knowledge and authority on farming to the educated youth. Through formalized education and training, traditional knowledge held by elders in farming communities may be valued less as young people are encouraged to embrace techno-scientific approaches to farming. The transition from farming as a way of life to farming as a business is contingent on the outcomes of this generational shift.

INTERLUDE: ON "MIXING"

I was first introduced to the idea of "mixing" when the elder Christopher Anabila Azaare, a self-trained historian, museum curator, climatologist, volunteer school teacher, and soccer referee, among other roles, used the term to describe the frequency of interreligious diversity within households in northern Ghana. When my husband, Anatoli Ignatov, and I met Chris Azaare at the end of 2011, he was just beginning to build a museum of Gurensi culture in the village of Gowrie, one cement bag at a time with minimal help. Chris and Anatoli—a fellow researcher studying land rights and indigenous knowledge—had a mutual interest in collecting local histories.[1] Chris had been obviously doing so for decades, as was evidenced by the enormous stacks of papers forming handwritten books that sat dusty on his desk. His genealogical maps of local families—delicately taped together and sprawling well beyond the parameters of the desk when unfolded— were treasures that settled both local curiosities and disputes. Chris was who you would speak to if you needed a question of family history or of culture answered, chiefs included.

Through working with Chris, my husband and I got to know the family of an influential chief in northern Ghana. Over the years that we visited the family, I got to take part in different religious celebrations—traditional, Christian, Muslim—with at least one member of the family celebrating such customs. Living in northern Ghana, we had friends with names that indicated familial shifts in religious identification over time—at times with Muslim, Christian, and traditional names fashioned together in one identity. We had friends that knew intimately the correct ritual practices and prohibitions for religious practices that were not their own. When we asked Chris about this apparent fluidity across religious practices despite religion's and tradition's own (competing) dogma and rules, he laughed, as Chris often does, and said, "Oh, they're mixers." "Mixing" is seen with the Muslim man that also pours libations on his family's ancestral grave, the Christian that is devoutly present at church every Sunday morning and Wednesday evening but also visits a soothsayer to know what animal sacrifice is necessary for their prayers to be answered, and the Catholic priest that also fasts with his Muslim coworkers during Ramadan. This fluidity across practices is not only in the borrowings and coexistence of different religious practices but also in the pragmatic blending of different agricultural practices and knowledges in northern Ghana.

Proponents and opponents of the new Green Revolution for Africa see the introduction of genetically modified crops and the expansion of commercial agriculture as *transformative*. However, attention to the commonplace practice of mixing suggests that such totalizing effects are not likely to pan out. What I found over the years studying efforts to promote a commercial seed sector in northern Ghana was that some of the model farmers that were supposed to lead the change in northern Ghana were themselves mixing: implementing and utilizing technological practices promoted by development planners while also, for example, practicing traditional labor-intensive weeding to minimize the application of chemical spray. Mixing is reflected in the endurance of seed sharing and saving practices in northern Ghana despite decades of efforts to try to increase adoption of improved seed. In interviews, this reluctance of farmers to adopt and to purchase commercial seed was a major source of frustration among development planners working on the seed sector, who viewed "seed sector reform" as a critical component of the transition to farming as a business.[2]

I first referenced mixing in the preface with the story of Bakari Nyari: an expert, entrepreneur, activist, community leader, bureaucrat, teacher, farmer, and mentor. Bakari navigates these different role performances while committed to the belief that the "communal nature" of land needs to be protected and has used his positions of authority to do so. In the two vignettes that follow, the practices of mixing of agricultural practices are described through the stories of lead participants in agricultural commercialization schemes that are a part of this new Green Revolution for Africa. Like Mr. Nyari—and despite their central position as actors within these efforts to commercialize and professionalize agriculture—they blend together concerns that go well beyond those immediately tied to commercial production for export: that is, food security for their laborers, environmental protection, biocultural diversity and ritual, and resource use.

"I Do My Part to Save the Environment"

I met Fatimata Adongo at the Mango Value Chain workshop that I described in the previous chapter—it was easy to approach Fatimata given her warm smile and her presence as one of the few other women in attendance at the event. Fatimata comes from a family that is well-known (her mother was an esteemed traditional healer), as is her family's mango operation that she manages. Further solidifying her rank as a serious farmer and a successful entrepreneur at the workshop is also her status as an Export Development and Agricultural Investment Fund grant recipient.

Whereas Fatimata's family business is meeting export demand with her 250-acre mango farm just outside of Tamale, she is one of the entrepreneurs that also recognizes the importance of local food security. While the mango farm requires the recruitment of labor, permanent workers farm crops for their family. During my first visit to her nursery, she told me that they had actually acquired more land than the 250 acres dedicated to mango production so that the farmers that work the farm were given land to grow crops of their own along the boundaries. This way, "the farmers see a direct benefit ... and have a stake in our success."

This approach, she explained, "generates more security." The farmers intercrop with groundnuts, beans and rice—"leguminous crops." They employ people in the surrounding communities, and this better standing in the community generates security.[3]

I was intrigued by her use of the word "security," as she used it to convey both food security and the security of her enterprise, viewing it as a mutual form of risk reduction. When I asked more about the land she had acquired for her workers' plots, she described how there is an expectation in her community that your relative wealth alone creates shared expectations that you are supposed to contribute to your community in whatever way you can. She has a social obligation to her workers but also to a large extended family as well as a sprawling network of her mother's patients and schoolchildren.[4]

She is an entrepreneur, but not *Homo oeconomicus*; she is not merely pursuing profit's bottom line; she considers herself embedded within a wider network of social and ecological relationships.[5]

Fatimata stressed that "providing food for families comes first," followed by the production of cash crops. This prioritization of family food security was a source of tension at the Mango Value Chain workshop, as organizers were trying to encourage farmers to get serious and reorder such priorities to focus on export production. Fatimata, among other farmers and officials managing agricultural projects in northern Ghana, told me it was a common practice among farmer participants in programs like Masara N'Arziki to distribute inputs such as fertilizer to the staple crops that would feed their families as well as the cash crops the inputs were intended for. This indicated a deviation from the formula and would likely lead to lower yields of the cash crops than expected. It also showed the privileged position of household provisioning that was maintained, despite efforts to recalibrate such priorities.

Fatimata not only "mixes" the logic of cash cropping for export with concerns about immediate food security needs; she also mixes farming practices. In addition to the mango farm, her family also runs a nursery of traditional medicinal plants and her mother practiced traditional healing. Fatimata explained to me that whereas the plants and trees used for healing are grown organically and abide by certain practices to maintain wild cultivars, "they also do commercial … use inputs—some chemicals" on their mango and cashew crops to "make it easier" and "improve upon yields." She explained to me that she "doesn't feel bad because in other ways … I do my part to save the environment."[6]

"If These Traditional Crops Are Displaced, How Will People Perform Their Libations?"

John Awedoba, briefly mentioned in the previous chapter, works for a northern Ghanaian nongovernmental development organization and has been selected as a lead farmer participant for a USAID Feed the Future farmer program that facilitates access to improved seed, agro-chemicals, fertilizer, and center pivot irrigation. He is based in Bolgatanga, in the Upper East region of Ghana, but grows irrigated maize, rice, and soy in Walewale, Northern Region, some 60 kilometers from where he lives. The NGO that John leads distributes hybrid and improved seed to farmers in the Upper East as part of its work, which he explained was to promote food security and income-generating activities in the Upper East.

The Upper East is commonly identified as the poorest region in Ghana, with much of the agricultural practices characterized as "subsistence-based" with rain-fed agriculture, low uptake of improved seed, and little to no consistent fertilizer application. It has some of the highest population density in Ghana, and the landscape is notable for its large deep reddish-brown boulders, baobab, shea, *dawadawa*, kapok, and whitethorn trees that intersperse between small farm plots not typically more than two acres, and many much smaller. Small dams, such as the one in Vea used to support agriculture in villages like Nyariga, have suffered from lack of infrastructure investment for decades. Notwithstanding local farmers' labor pooling to remove the excessive silt in the dam, the irrigated water from the dam fails to travel far—leaving more farmers dependent on rains that are far more erratic and destructive than they once were.

Speaking to John was particularly interesting because he is both a participant in and promoter of improved seed in both the Northern and Upper East Regions via his participation in the Feed the Future program and his work, respectively, but he is also highly critical of these interventions in the seed sector. He expressed cynicism regarding the starter packets offered to farmers as part of the Feed the Future program, stating that what this did was create dependency on inputs and create new markets for foreign businesses. I asked John Awedoba if the USAID Feed the Future crop focus on maize, rice, and soybean was appropriate for the Upper East Region. John responded, "Yes and no." He explained that while the sale of these cash crops brings in income, maize and rice are "high feeder crops," meaning that they require high levels of applications of fertilizers. By contrast, traditional crops such as millet, sorghum, cowpea, or Bambara beans are "low feeder crops" that require little inputs. He summed up that improved seed required more inputs and ultimately creates new demand for fertilizer and agro-chemicals.

John Awedoba continued in his assessment of Feed the Future,

> Feed the Future is more concerned with "food security" while I am concerned with "food sovereignty." These crops could displace indigenous seed. If these traditional crops are displaced, how will people perform their libations? How will they perform their rituals—funerals and wedding? People don't just grow crops for food—there's this cultural importance to traditional crops.[7]

He went on to say that he is concerned that these interventions in the seed sector could impact local, traditional seed breeding, suggesting that such practices could be displaced by seed companies. John mixes the work of promoting these seed sector interventions while, on a personal level, advocating for traditional crops. He values food sovereignty but works for food security.[8] He worries about the potential displacing effects of these changes to the seed sector, yet he participates in, and even advocates for, these alterations to agricultural production from the cultivation of crops that require little to those that demand a lot. John is a "lead farmer" entrepreneur who is supposed to get other farmers to emulate the cultivation practices that will "feed the future" and commercialize agriculture in Ghana. But John is also a "mixer" concerned with the "cultural importance" of traditional crops and farmer indebtedness related to this shift away from subsistence.

Mixing and the New Green Revolution for Africa

Mixing implements certain aspects of commercialization and professionalization of farming but also values and seeks to protect traditional practices. Mixing recognizes that diversified livelihood strategies provide buffers to shocks in food prices and climate crises by planting food for the family *first* and then cash crops only thereafter. Mixing is seen in the commonplace practice of farmers trying out seed and saving the ones they like, often for many generations and in community exchanges to diversify what farmers can grow. It furthermore reflects what Scoones (2009: 172) describes as the way in which people "combine different activities in a complex bricolage or portfolio of activities" rather than in the neatly defined categories typically used to characterize rural development ("agriculture, wage employment, farm labor, small-scale enterprise" (ibid.)).

Mixers reject dogma and monoculture (or in the case of religious mixers described earlier, monotheism), which makes it less likely that efforts to modernize agriculture as part of the "new Green Revolution" in Africa will fully take root. Mixing resonates with the experimentation, flexibility, and pragmatism that Richards (1985) and Berry (1993) highlight as characteristic of West African farming practices. As one female scientist told me, "Rural women experiment everyday. They keep talking to each other; they do a lot more research than we think they do. They keep experimenting."[9] Additionally, as Richards (1985: 145) shows, "Farmers may prefer the variety inherent in local planting material to the uniformity of pure line selections … this preference reflects peasant concern to keep open as many options as possible in the face of environmental uncertainty." Mixing is understanding that your great-grandfather could have as much to say about how to farm as would development planners with improved seed: mixers judge the comparative value of each by observation and experimentation, as well as tradition. They are innovators of technology, not merely users. Pragmatic "people's science" guides their farming decisions as they stay open to new, promising opportunities to resolve the challenges that they face and to improve their livelihoods (Richards 1985; Berry 1993; Scoones 2009, 2015; Scott 1998: 264).

The practice of pragmatic mixing challenges efforts to transform farming from a way of life to a business in Ghana as farmers pick and choose what kinds of "modern" or "traditional" agricultural practices to adopt. That is to say that they hold multiple roles that secure not only income but also food for their families and seed commons for their future. These two modalities of farming are not seen as opposing or mutually exclusive but rather hybridized and bundled within one another. Flexibility is a source of resilience not only pertinent to decisions around which crop varietals to cultivate (Berry 1993: 199) but also reflected in the multiple sources of income streams and employment opportunities pursued as part of diverse livelihoods strategy (Scoones 2015, 2009). Entrepreneurship in this context is not about a singular occupation but, rather, part of a diversified livelihood that balances concerns not only about income but also about the food security needs of workers and extended family, the relationship to land and culture, and environmental stewardship. Such practices of mixing complicate both the agricultural and identity transformations that are crucial components in this new Green Revolution for Africa and challenge the idea that this agenda will be as transformative or unidirectional as both proponents and opponents claim.

Chapter 6

NEOCOLONIAL ANXIETIES

I. Introduction

On May 4, 2015, I was walking through Asylum Down when I stumbled upon the headquarters of the Convention People's Party (CPP), the party of Kwame Nkrumah, now one of the smaller political parties in Ghana. I had planned to interview members of the CPP given their opposition to the entry of genetically modified organisms (GMOs) in Ghana, but I had only just returned to Accra the night before. I do not like missed opportunities, so I entered the headquarters and introduced myself as a researcher studying the politics of GMOs in Ghana. I was in luck—one of the spokesmen for the CPP had "just returned from court" and I could speak with him. He was energized: the CPP had just joined the lawsuit that Food Sovereignty Ghana initiated against the Ghanaian Ministry of Food and Agriculture and the newly formed National Biosafety Authority for violations of the Ghana Biosafety Act of 2011. I asked why the CPP, in particular, is opposed to GMOs. He responded, "We see GMO as neocolonization. We don't want something imposed." He added that "Kwame Nkrumah, the founder of our party, had written about neocolonization … we see GMO as a manifestation of his prophecy" with "GMOs making their way into the country" by way of new supermarkets—including some from South Africa—carrying GMO products.

"In the end, we don't want [GMO] at all because we don't need it at all." The youth spokesman made this point by arguing about the amount of postharvest crop loss that the country experiences: "farmers are already in bad condition" with "thirty percent of the harvest spoiling" due to bad roads or a lack of storage. He elided the suggestion that genetically modified seeds would hold any advantages to Ghanaian farmers, "GMO seeds are so expensive and farmers are forced to buy company chemicals … Ghana will be developing their economies, instead of ours." Further, he elaborated that the loans for farmers to purchase GMOs may be a problem if the yield goes down and they are left with debt. Then, "farmers will be kicked off farms when they don't do well." When I pushed for greater clarity about how this kind of "land grab" would unfold, he walked back his statement and stated that when farmers don't do well "they will sell," citing southern Nigeria and India as examples. But these "economic effects" are not appreciated by the "GMO pushers or lobbyists who are concerned with how they can sell GMO seed and chemicals." The spokesman emphasized how the concern about GMOs elsewhere warrants a cautious approach, "If big countries like France ban GMOs for health reasons, why should we go? It should be a food democracy, but America says no."

In the time since December 2011 when I began my research studying the politics of agricultural modernization and genetically modified crops in Ghana, my informants shared with me—repeatedly—concerns about the role of outsider influence in contemporary deliberation over the future of food and agriculture within Ghana. Agricultural research scientists complained of the impact of European-influenced "anti-groups" that were creating hostile conditions for the development of transgenic crops for public benefit. An official of the Ministry of Food and Agriculture commented in a meeting about "covert coercion" by USAID to get Ghana to accept genetically modified crops. An agricultural research scientist frustrated that the *Bt* cotton trials were abruptly ended in Ghana blamed France's sway on the Burkinabe government for legislation against Monsanto, "France is tightening its colonial grip." This chapter explores such emerging discourses among Ghanaian scientists, policymakers, and activists that view Ghana's agricultural policy through the lens of neocolonialism.

In considering the contemporary challenges to food sovereignty in Ghana, the chapter considers Nkrumah's understanding of neocolonialism, a view that shapes modern interpretations of the phenomenon given the reach of his pan-African philosophy: "The essence of neo-colonialism is that the State which is subject to it is, in theory, independent and has all the outward trappings of international sovereignty. In reality its economic system and thus its political policy is directed from outside" (Nkrumah 1965: ix). In *Neo-Colonialism: The Last Stage of Imperialism*, Nkrumah articulates various "mechanisms of colonialism," that is, "modern attempts to perpetuate colonialism while at the same time talking about 'freedom'" ([1965] 1987: 239). He specifies that these mechanisms include indebtedness, development "aid" and its associated conditions, capital's control over the world market and trade agreements that disadvantage developing economies, and donor restrictions that require recipients to use funds in particular ways such as to purchase goods from the donor country (Nkrumah [1965] 1987: 239–43). Nkrumah is attentive to how neocolonialism operates not only in the economic sphere but also politically, ideologically, and culturally by way of the development industry, militarism, and propaganda that reproduce the disenfranchisement of African states (239).

Building off of Nkrumah's "mechanisms of neocolonialism," the chapter uses the language of "vectors" of neocolonialism to consider how neocolonialism may spread not only through economy and law but also through ecological means. The meaning of "vector" refers not only to the transmission of a pathogen by an organism or its use in genetic engineering to move genes across species (Shiva 1999: 33) but also to magnitude and direction. As Shiva (1999: 63) describes in her critique of genetic engineering, despite biotech advocates' presentation of genetic engineering as imbued with precision, there is a significant lack of control in the ultimate outcome of the genetically modified organism due to an inability to target the vector at specific sites within the recipient genomes. Therefore, to speak of vectors—rather than mechanisms—creates space for contingency and changes in magnitude and direction.[1] For example, if the state were to embrace food sovereignty, this information would affect change in these vectors (reducing their magnitude or altering their direction or both).

In the following sections, the chapter explores three different "vectors of neocolonialism": ecological, economic, and legal. We can bear witness to ecological vectors of neocoloniality by appreciating the forces set in motion with the "invasion" of the fall armyworm (FAW) in Africa as well as heightened ecological vulnerability experienced by those on the climate frontlines of northern Ghana. When considering the "economic vectors of neocolonialism," I look to an effect of structural adjustment: the slow violence (Nixon 2011) of weakening institutions of plant research and breeding. These conditions incentivize public–private partnerships (PPPs) that, rather than producing more equitable and pro-poor relations (Morvaridi 2012), enable outsized donor influence on agricultural priorities in Ghana given both the potential resources and the prerequisite "enabling environment" that has to be fostered to attract participation. Legal reforms in the domains of plant breeding, intellectual property rights, and biosafety suggest that legislative change's greatest beneficiaries may be global agribusiness; this vector of neocolonialism has the potential to create new economic relationships (and dependencies) to global agribusiness in the seed, agro-chemical, and food markets and creates the conditions for biocapital. Important to this examination of contemporary neocolonial politics is a critical consideration of what food sovereignty—defined here as the ability for nations and communities to determine how and what food should be produced, distributed, and consumed—looks like in the Ghanaian context. The chapter concludes by drawing upon Agarwal (2014) to critically analyze how this discourse of neocolonialism illuminates tensions between democratic decision-making and popular notions of food sovereignty.

II. Ecological Vectors of Neocolonialism: The "Invasion" of the Fall Armyworm in Africa

Nkrumah, while alert to neocolonialism in many of its forms, was less attentive to its ecological expressions beyond concern about monocropping's impacts on food crop diversity and prices for African farmers. Yet, as I will highlight in this section, the changing political ecology of agriculture in Ghana creates conditions for neocolonial influence that is epistemic as well as material in nature.

The insect known as the invasive fall armyworm (FAW) arrived in West Africa in 2016, likely traveling via imported maize from the Americas, where it is native (Briggs 2017). As written in a 2017 *Nature* article, "Although no one knows how the insect got to Africa, increased trade and climate change are the likely culprits" (Wild 2017: 14). A *Current Science* article dramatized concerns regarding a "fall armyworm invasion": "Fall armyworm (*Spodoptera frugiperda*) has already invaded almost half of Africa since its first observation in the continent in January 2016. … At its current rate of invasion, the pest may conquer Africa before the end of 2017" (Nonzom and Sumbali 2018: 27). Its first arrival in Ghana was documented in the Eastern Region in April 2016, and by June 2017, a total of 120,000 hectares were affected, accounting for destruction of 1.6 percent of the production area of maize in the country.[2] Between 2016 and 2017, the farmland destroyed by the fall armyworm increased from 4,046 hectares to 14,411 hectares, with maize losses alone—a major staple crop in Ghana—accounting for 27,669 metric tons in Ghana (ibid.). Although the insect consumes a range of 80 different crops, including cowpea and sorghum, it has a distinct preference for maize. Ghanaians that farm at a distance, as is common for urban residents, arrived to find plots devastated. One informant told me that in the two weeks that it took to return to his field, it was decimated: "last year I didn't even get a grain of maize."

Awareness campaigns in the form of jingles that played on 43 stations across the country, glossy posters of the insect magnified with warnings legible from across a room, as well as trial-and-error applications of pesticides and homemade concoctions[3] formed the initial response to FAW's arrival in Ghana.[4] During the 2017 Joint Sector Review (JSR) meeting in Accra, the agricultural minister for the Plant Protection and Regulatory Services Directorate highlighted the importance of pest surveillance and research on biological controls; however, great emphasis was placed on the need for a "strategic stock of pesticides" with a recommendation that "inputs for Planting for Food and Jobs should include insecticides."[5] Sisay et al. (2018: 801) report, "Invasion of fall armyworm in Africa alarmed governments of different countries in Africa to apply different insecticides in maize fields as an emergency response to fall armyworm invasion," consistent with the accounts from my interviews with the Ministry of Food and Agriculture officials and attendance at the 2017 JSR meeting. At a question-and-answer session during the JSR meeting, a representative of the UN Food and Agriculture Organization made a point of how little attention there was on the public health and ecological effects of the "experimental" combinations of chemicals used to combat fall armyworm.

Integrated pest management strategies that include scouting and surveillance as well as increasing predation of these insects (through biocontrols of parasitic insects) are among the strategies advocated for by international agroecologists and biotechnologists alike, with the use of chemicals generally advocated as a last resort. This approach informs the FAO and the Centre for Agriculture and Bioscience International (CABI) training of trainers' manual to support community-based fall armyworm monitoring, scouting, and early warning system implemented by the donor community[6] (FAO and CABI 2019). Dr. Joseph Huesing, entomologist and a former manager of Monsanto's Regulatory Scientific Affairs division, has been a lead expert on responses to the FAW in Africa, emphasizing the role of biotechnology in responding to this agricultural threat. In his 2018 presentation, "Fall Army Worm in Africa: A Guide for Integrated Pest Management," the senior biotechnology advisor for USAID (and coauthor of a guide bearing the same name) states that fall armyworm is a "challenging pest to control" and that "FAW will be endemic in much of Africa—it's not going away." Huesing then shifts to a solution trifecta, "Tools—Knowledge—Policy," stressing that "tools require an enabling policy environment." Huesing and his coauthors argue that the approach to combatting the insect "should be informed by sound scientific evidence, build on past experience combatting FAW in other parts of the world, and be adaptable across a wide range of African contexts (particularly for low-resource smallholders)" (Prasanna, Huesing, Eddy, and Peschke 2018: 6). The logic of learning from past experiences turns attention to the experience of Brazil and the United States, where FAW is native and "~85% of farmers choose GM maize for control" (Huesing 2018).

A "study tour" to Brazil designed to teach African policymakers about responses to fall armyworm was organized by Africa Lead, a capacity-building program under the umbrella of the US government's Feed the Future food security initiative. The study tour and engagement with Brazilian FAW control authorities sought to both expose African decision-makers to proven and "successful" technology such as the use of genetically modified crops and encourage African policymakers to create the enabling environment required to develop or apply these technologies in their own countries (Africa Lead Team II 2018: 49, 35). This further underscores that an enabling environment of legal reform, market reform, and monoculture promotion is a prerequisite for the introduction of genetically modified crops and has the effect of increasing demand not only for the technology but for experts versed in the technicalities of intellectual property law, commercial plant breeding, and biosafety regulation.

As with other invasive species, fall armyworm is able to proliferate in part because of less native predators and a general lack of awareness of an insect that has been in Ghana only since 2016. Maize monocultures foster the insect's expansive reach and appetite. My Ghanaian friend, a climatologist based in Nyankpala, was one of the farmers whose maize plots were decimated the first year that FAW made its arrival in Ghana. "The heat is making the pest infestations worse because the cycles are sped up." In his view, the threat of FAW was exacerbated by climate change: warming weather meant ripe conditions for their reproduction and consequential destruction of fields. Here the twin threats of climate change and invasive species converge to create new ecological threats for food and livelihood security on the African continent.

The injustice of climate change, whereby those peoples who have contributed the least to this global problem bear its greatest burden, at once acutely affects those Ghanaian farmers dependent on rain-fed agriculture or those maize farmers hit by a fall armyworm infestation whose devastation is accelerated by warming weather. Such ecological disruptions heighten foreign influence over African agricultural systems as experts and technologies from the Americas are turned to deal with invasive pests from which they came. Hype over GMOs and "climate-smart" agriculture, discussed in Chapter 4, is intensified as the impact of droughts and pest infestations affects frontline communities across the African continent. I argue that neocolonialism is expressed not only by economic, political, and legal conditions that extend the influence of powerful foreign actors but also by the ecological changes that indirectly maintain its influence.

III. Economic Vectors of Neocolonialism: State Capacity and Donor-Driven Development

The narrative of an excluded Africa, described in Chapter 1, presumes growth is contingent on the delivery of technical products from the West. This narrative creates demand for foreign funding and expertise to fill the gaps in African research and development that structural adjustment left behind. Austerity measures in African countries that slashed public-sector expenditures for plant breeding and other agricultural expenditures (Chambers et al. 2014: 28; Gibbon 1992: 59) and led to a dramatic decline in donor funding for agricultural research and development in the mid-1980s (Spielman, Zaidi, and Flaherty 2011) created massive deficiencies in the coverage of agricultural extension. This austerity-induced economic crisis also led to a mass exodus of qualified staff from African countries, undermining the capacity of the national agricultural research institutes to conduct research, impacting their research agendas as well as the morale of agricultural research scientists (Puplampu and Essegbey 2004: 275, 277; Mkandawire and Soludo 1999: 135).

This narrative of an Africa excluded from technological revolutions creates the rationale for development interventions with a technological focus, such as the USAID-funded Agriculture Technology Transfer (ATT) Project and US bilateral assistance in support of agricultural biotechnology. Relatedly, the Alliance for a Green Revolution in Africa, the World Bank, and the Syngenta Foundation[7] fund research and plant breeding for national agricultural research institutes and new educational institutions and programs that include the West African Centre for Crop Improvement (WACCI), the MS degree in Seed Science and Technology, and the MS degree in Genetics and Plant Breeding at the University of Ghana, Legon, and internships with Dow AgroSciences and Monsanto as part of these degree programs.[8] The support from these actors is not without consequence: as Puplampu and Essegbey (2004: 279–80) argue, the lack of state capacity to substantially fund agricultural research "accounts for the donor-driven nature" of research in the region as African governments rely on PPPs to fill gaps to compensate for their diminished ability to provide public goods. As argued in Chapter 2, how donors view problems and priorities may not align with the views of governments and citizens, and this can lead to projects that do not reflect recipients' interests.

Furthermore, donor funding may also require recipient countries to support these initiatives (financially or through policy change) as part of the PPP framework (McGoey 2014). That is, a certain amount of state capacity—an enabling environment, if you will—is needed by states to attract donor and private support. We should expect declines in philanthrocapitalist support when there is not adequate state support given that it is states, not the private sector, that are risk-takers (McGoey 2014: 122). It is therefore unsurprising that once African states committed to increase spending on agricultural research and development to a minimum of 10 percent of their national budget in 2003, private sector investment, foundation funding, and bilateral aid increased (Fan, Omilola, and Lambert 2009). The African Union's Comprehensive Africa

Agriculture Development Programme (CAADP) Compact, which Ghana signed in 2009, encourages this state-level agricultural spending used for "improving food and nutrition security, and increasing incomes in Africa's largely farming-based economies. It aims to achieve this by raising agricultural productivity and increasing public investment in agriculture" (NEPAD n.d.). This commitment to prioritizing agricultural development is further demonstrated by Ghana's participation in the "Grow Africa" partnership, co-convened by the African Union Commission, the New Partnership for Africa's Development Agency (NEPAD), and the World Economic Forum.

An analysis of Ghana's "agriculture transformation" agenda by the African Union's NEPAD tells us that value chain development by way of public–private partnerships, access to agricultural inputs, improving food security and nutrition, and completion of the CAADP process to "harmonize" institutional conditions are on track to meet their CAADP implementation goals (NEPAD 2018). Evidence of meeting these goals include PPPs focused on the uptake of agricultural technologies; increased spending on supplemental soy nutrition in schools and in university research collaborations to develop a commercial soy market in Ghana; and the legislative changes that standardize plant breeder protections, intellectual property rights, and biosafety protocols. The trend anticipated by McGoey (2015, 2014) of increased private sector investment following increased public sector investment in agriculture is demonstrated in the Feed the Future Agriculture Technology Transfer Project (discussed in the previous chapter) briefing presentation, "Progress in numbers." The 2018 brief shows that the ATT Project was able to increase the sales of "targeted ATT commodities" of seed, fertilizer, and soil amendments by 789 percent, whereas the "value of new private sector investment in the agriculture sector of food chain leveraged by FTF implementation" grew by 168 percent through US bilateral assistance (Feed the Future n.d.).[9] Feed the Future implementation, CAADP harmonization, and philanthrocapitalist PPPs are generating private sector investment and expanded markets for these commodities. This is consistent with the expressed goals of the US State Department Office of Agricultural Policy (2019) that commits to advancing US national interests by "opening foreign markets to American products" and using global food security programs like Feed the Future "[to support] U.S. national security," the latter understood here as advancing American business interests. These dynamics speak to Nkrumah's concerns of how development assistance can be neocolonialist when used to push policies that lack popular support or encourage recipients to "buy goods from the donor nation" rather than develop local industry (Nkrumah [1965] 1987: 243).

As I further develop in the following section on legislative change, these efforts to create an enabling environment for investment and commercial agricultural success beg the question of *whose success*. There is no doubt that there will be Ghanaian farmers, agro-dealers, plant breeders, experts, and intellectuals who will benefit from the returns on the agricultural transformation agenda currently underway. But will these benefits extend beyond the elites? How are the gains distributed? Will this create the conditions for Africans to solve their own problems, or will this create new needs for foreign expertise? What kind of economies and markets are being fostered? What kind of food security is being promoted? In the following section I consider *who benefits* from Ghanaian legislative reform in plant breeding and intellectual property rights.

IV. *Legal Vectors of Neocolonialism: Legal Reform in Plant Breeding and the Seed Sector*

Let us return to the Plant Breeders' Bill (PBB)[10] that has been subjected to severe criticism by the food sovereignty movement in Ghana and discussed in Chapter 3. Plant breeders' protections are presented by proponents as necessary for the future of food security; "the introduction of a legal framework to support the protection of the rights of breeders of new varieties … is of paramount importance at this point in time when the food situation on the globe is not only precarious but uncertain as the world population continues to grow" (Republic of Ghana 2013: 1). Its advocates presume that the PBB will help the country increase yields and compete in international markets. Such claims are treated with cynicism by not only food sovereignty activists but also media that, at times, has reported on this legislation with skepticism about the true motivations behind the legislation.

So, what's at stake with the passage of such legislation? Aistara's (2012) study of the imposition of intellectual property rights on seeds in Costa Rica helps to illustrate this. The contested legal reform—for Costa Rica to become compliant with the 1991 Union for Plant Variety Protection (UPOV) Convention as part of its conditions to join the Central America Free Trade Agreement—creates what she terms "privately public seeds." This happens when those that qualify as plant breeders are free to access protected varieties for research, whereas new restrictions on access exist for a substantial number of farmers who do not have the means to produce the genetic stability and uniformity to qualify as breeders (Aistara 2012: 134). She counters arguments that emphasize how this will lead to the privatization of seed, arguing that the more pronounced effect is rather the privatization of knowledge:

> It is not the seeds themselves that will get privatized, but perhaps more importantly, it is the knowledge surrounding seeds and the privilege of seed management that get appropriated through drawing a distinction between breeders and farmers. (133)

The PBB, now called the Plant Variety Protection Bill, in essence, is part of the Economic Community of West African States' (ECOWAS) seed harmonization agenda to align themselves with the UPOV. On page one of the bill it also explicitly states that it "permits farmers to save and replant seed and provides them with the right to use protected varieties as a source of further research and breeding activities." This latter element stands in tension with the intellectual property rights that are also being promised protection through this regional seed harmonization, part of a global shift in the standardization of legal protections on plants, seeds, and intellectual property that Costa Rica is also a part of. Although different than Costa Rica in important ways, we can expect that if there is not equal access to the rights of use and information on seeds (by cutting off farmers from access because they are not viewed as "breeders"), we will see uneven development in terms of research and innovation. Such arrangements advantage breeders and institutions with capital that can meet the standardized criteria (seeds with uniformity and stability, often developed under the controlled conditions of laboratories) required for UPOV protection and enables the exclusion of farmers that do not appear to be engaged in "research" or "breeding."

When discussing plant breeders' rights with one representative of a peasant farmers' association, he stated, "UPOV is colonial. ... It is a neocolonialist law."[11] Food Sovereignty Ghana also views such legislation to be against the interests of Ghana and its farmers:

> The giant multinational corporations waiting in the wings stand to benefit at our expense. ... African countries in concert with other developing countries ensured during the TRIPS negotiations that a developing economy like Ghana would not be short-changed by giant corporate interests. The inclusion of the *sui generis* clause in the WTO negotiations was an important victory won by the so-called third world countries like Ghana. ... To consolidate the victory won at the WTO, the African Union developed a model that carefully took into consideration, the legitimate rights of the plant breeder, as well as those of our farmers. It is thus truly pathetic that the report submitted to Parliament made no mention of this. (Food Sovereignty Ghana 2018)

Food Sovereignty Ghana is referencing the African Union's work toward a common platform for the implementation of the International Treaty on Plant Genetic Resources for Food and Agriculture, which explicitly supports farmers' rights in Article 9 (FAO 1983: 12).[12] Equitable and improved accessibility to different varieties of seeds and enabling farmers "the equitable sharing of the benefits arising from the use of material in the International Treaty's Multilateral System of Access and Benefit-sharing, which currently includes over 1.4 million samples of plant genetic material for food and agriculture from around the world" were high-priority items at a 2017 African Union Commission workshop focused on the implementation of the International Treaty on Plant Genetic Resources (FAO 2017).

Joining Food Sovereignty Ghana in the critique of the Plant Breeders' Bill, the Peasant Farmers' Association of Ghana stated to *Daily Graphic* (February 2, 2017) that "the bill did not recognise the rights of indigenous seed growers and local farmers, the right to plant and replant seeds, the right to save seeds and the right to market and share seeds" and "the bill was only protecting the interest of foreign merchants and corporations."[13] The accounts by Peasant Farmers' Association of Ghana and Food Sovereignty Ghana are somewhat misleading given that the bill reflects a *sui generis* system that specifies that "the plant breeders' rights system permits farmers to save and replant seed and provides them with the right to use protected varieties as a source of further research and breeding activities." However, this is provided for in the context of *plant breeders' rights* and, as previously argued, not all farmers would meet the criteria for this recognition as "research and breeding."

What is certainly accurate about these accounts is that there is *no mention of farmers' rights*, and in the six times "farmers"[14] are mentioned in the Plant Breeders' Bill, half of the time they are mentioned as recipients of technology rather than generators of knowledge and plant breeding activity. Of note, the PBB was "drafted by the Attorney General's office" rather than the Ministry of Food and Agriculture "because it deals with IP [intellectual property]."[15] The Plant Variety Protection Bill differs from the PBB in name only, and as it stands (as of December 2020) would strengthen the rights of foreign and certain domestic plant breeders. As Manu (2016: 25) articulates, "It appears that the Parliament intends to allow the granting of legal protection that will arguably protect the rights of scientists and private corporations seeking to develop and commercialize GMOs in native species of seeds and crops—a system that historically has not been the subject of legal protection."[16] As discussed in Chapter 3, Clause 23 of the PBB specifies that subsequent Ghanaian legislation could not override these rights.[17]

While rejecting these kind of plant breeders' protections, Food Sovereignty Ghana has used the failure of the government of Ghana to implement the protocol established in the Ghana Biosafety Act as the basis of their first lawsuit. This move, although an understandable strategy to delay further development of GM crops,[18] reaffirms the legitimacy of this biosafety legislation that allowed GMOs into the country. Perhaps a more potent strategy would be to highlight the limits of informed consent in the preparation of the Ghana Biosafety Act, particularly the lack thereof secured in northern Ghana, as well as the level of lobbying parliamentarians were exposed to in the run-up to the vote that passed the Ghana Biosafety Bill (Rock 2018). The concerted effort of USAID's Program for Biosafety Systems, the Open Forum on Agricultural Biotechnology, the Forum for Agricultural Research for Africa, and the Cornell Alliance for Science to push for the Ghana Biosafety Act to be passed suggest the "covert coercion" one of my Ministry of Food and Agriculture informants articulated concern over. Manu (2016: 37) argues that the key intellectual property provision of the US African Growth and Opportunity Act has pressured African lawmakers to produce plant breeder rights' legislation as a condition for continued trade concessions, suggesting that the coercion may not be all that covert.

Another outside actor influencing policy change in Ghana's agricultural sector is the Alliance for a Green Revolution in Africa. As discussed in Chapter 3, AGRA's seed sector division of the Policy and Advocacy Program worked with the Ministry of Food and Agriculture to develop a national seed law in line with the 2008 ECOWAS seed harmonization regulation, Ghana's Plants and Fertilizer Act that created opportunities for private sector seed production (World Bank 2012b: 3). This suggests an entry point not only for biocapital but also for neocolonialism if this leads to the dominance of global agribusiness in these emergent private seed and agro-chemical sectors in Ghana. The PBB creates enhanced opportunities for private investment and the extension of plant breeders' rights onto foreign legal entities in the seed sector (Ignatova 2017; Ayenan and Danquah 2015: 1). In sum, the extension of plant breeders' rights to foreign legal entities, the narrow definition of plant breeders' rights reflected in the PBB and Plant Variety Protection Bill, and the shift initiated by the Ghana Plants and Fertilizer Act to allow private seed companies to produce foundation seed enable the foreign private sector to possibly dominate the Ghanaian seed sector in the future.[19] These legal reforms being ushered in from different directions create an enabling environment for investment, but the question still stands as to *who* is being invested in and *whose rights* are being protected.

V. Food Sovereignty

Food sovereignty is envisioned by many as a form of independence from such external influence, appealing to the postcolonial aspirations that Nkrumah and Sankara spoke of (Sankara, cited in Prarie 2007: 104). Initially framed by the transnational peasants' movement La Via Campesina in 1996 as "the right of each nation to maintain and develop its own capacity to produce its basic foods respecting cultural and productive diversity," food sovereignty has moved away from the nation-state-based definition to consider food sovereignty for peoples as well as communities. However, there is a lack of consensus on how to address differences in how food sovereignty is defined within communities, what the process would be in negotiating those differences and articulating policies, the fact that not all communities can be (or would find it desirable to be) food self-sufficient, where small-scale export-oriented farmers are situated within this paradigm, how much extra-local trade would be acceptable, not to mention desires for non-native products like coffee, spices, and spirits that have become integrated within local diets (Agarwal 2014; Edelman 2014: 973–4). An additional challenge to food sovereignty is climate change, which threatens frontline communities with harvest-devastating severe storms, drought, salinization, and flooding that undermine the attainment of local food self-sufficiency.[20] The discourse of food sovereignty is, indeed, a powerful mobilizing frame but leaves many key issues of its implementation unsettled (Agarwal 2014).

An underemphasized aspect of food sovereignty is how the devaluation of traditional knowledge over time shapes how peoples and communities understand their interests. That is to say, the promotion of techno-scientific approaches to agriculture in schools and agricultural research institutes may affect how peoples "define their own food and agriculture systems" over time. As Hountondji (1997b: 1–2) articulates:

> The fact bears repeating: in the fields of science and technology, Third World countries, especially those in Black Africa, are tied hand and foot to the apron strings of the West. ... Where, for instance, does all of the equipment used for

research come from? How are research topics selected? ... Who in reality are the intended beneficiaries of this research? ... How does it fit into the society producing it? And to what extent is this society able to take charge of its findings?

Such a statement resonates with my observations at the Savannah Agricultural Research Institute, where donor funding opportunities appeared to shape what areas of research were pursued, and how. These dynamics of funder influence on research trajectories run parallel with the ways by which local farmer experimentation is given insufficient attention and support, despite, as one of my research scientist informants put it, "rural women experiment everyday."[21] She continued by saying that it was commonplace for members of the scientific community to bring their ideas to farming communities but not to listen, despite the active role that these farmers play in experimenting and in trying to solve their own agrarian problems. The idea that the scientific community has little to learn from local farmers was restated in various ways in my years of research in Ghana. When I asked one representative of a farmers' association why he thought farmers continued to save seed rather than purchase "improved seed," he stated that it was "the system they are used to" and that is why it was important to use demonstration fields to introduce new technologies as "they are like children."[22] At the JSR meeting I attended in Accra in 2017, farmers were generally talked about as if they just needed to be told what to do, a blank slate if you will, without reasons and rationale for their own farming and risk-reduction practices.[23]

Food sovereignty is a profound *challenge* to this idea of farmers as ignorant figures who need to be taught; it reasserts the importance of farmers' traditional and indigenous knowledge as well as the right to define their own food and agricultural systems. Take, for example, the Centre for Indigenous Knowledge and Organizational Development, who have been at the forefront of food sovereignty activism in Ghana since they were registered as a nonprofit in 2003. They approach their work to support rural livelihoods through an endogenous development approach that draws from traditional and indigenous knowledge in order to "ensure that our work is community driven and that we are building the capacity of communities" (CIKOD n.d.). As Bern Guri, executive director of CIKOD, articulates,

The trend these days is for African leaders to go seeking external interventions to solve our African problems. The fact is that over the decades of external investments Africa is still hungry. ... But the message I want to give is that if Africa is to solve its food problems we must begin to look for African solutions. ... This means, let's start with what our local farmers have and what we know best ... the farming systems that our local farmers know best and have used to feed us all these years, the seeds that they have used all these years, the soil management methods that we have depended all these years, and then we can go ... to ask for resources to scale up what we have developed ourselves. ... In other words, what we Africans should be advocating for is food sovereignty and not food security!

In their work on rural development, food sovereignty is advanced when local communities' existing agrarian practices are supported and communities have the ability to define their own food and agricultural systems.

A focal point of the food sovereignty movement in Ghana has been anti-GMO resistance. This resistance in Ghana has been largely concentrated on drawing attention to the presence of GMOs in Ghana, warning about the technology's hazards as they view them, organizing local March Against Monsanto demonstrations and other protests, and fighting plant breeders' protections that disenfranchise smallholder farmers. The Ghanaian food sovereignty movement was not able to stop the passage of the bill that allowed GMOs into the country—Ghana Biosafety Act—as the bill passed quietly following the USAID Program for Biosafety Systems "awareness creation" and legislative advice.[24] It was then that the Coalition for Farmers' Rights and Advocacy Against GMOs (COFAM) was formed, which includes Food Sovereignty Ghana and the Centre for Indigenous Knowledge and Organizational Development, leading a campaign of workshops on food sovereignty, organizing protests, and disseminating information against the PBB. Food sovereignty actors are transnationally connected, sharing information and amplifying their message through connecting with like organizations such as the African Centre for Biodiversity, which has been at the forefront of resisting genetically modified crops in South Africa.

Protests are not the only sites where such criticisms emerge about the outsized influence that foreign actors have on the politics and policy of food and agriculture in Ghana. Agricultural research scientists speak of the role of these "anti-groups," as they call them, and argue that this opposition to biotechnology is based in misunderstanding: "The idea is that we want the media to understand what are the *facts* [stress on this word] because we know, we recognize the fact that the 'anti-groups' are out there and most of them have contact with those [food sovereignty] groups."[25] Another agricultural research scientist expressed frustration that biotechnology was not more openly embraced in Ghana, attributing this to NGOs "polluting the minds of farmers."[26] When I interviewed the Ministry of Food and Agriculture's Board of Directors in March 2013, I asked board members about the benefits and concerns regarding GM crops. My questions sparked a lively debate. Some board members expressed positions in favor of GM crop cultivation: "agriculture is confronted with lots of issues biotic and abiotic, we need innovative ways to address them"; with biotechnology it is possible to "increase yields," "address yield gaps," and "make crops able to deal with the environment." Others emphasized that it would be more appropriate if biotechnology were "home-bred" with "Monsanto and Syngenta [linked] up with our varieties." Yet in that room at the Ministry of Food and Agriculture, much like the activists who were criticizing the Ghanaian government, there was a clear concern about the appropriateness of GM technology, its impact on seed saving, and the dependency that GM seed presents, both "financial and material." When talking about implementation of biotechnology and biosafety in Ghana, one particularly influential board member stated, "We don't want covert compulsion ... for some time we have seen it, and we don't want that. We know that Monsanto and Syngenta funds a lot of research. We need to agree to a course of action."[27]

Activists and bureaucrats, which may have many points of contention regarding the desirability of commercial agriculture, are actually in agreement in demanding greater democratic participation in food policy—and not "coercion." Yet this idea of democracy may be also taken to mean consideration of the introduction of GM crops, if the latter were understood as one way of diversifying the options on the table, rather than the major viable possibility. One of the agricultural research scientists at SARI characterizes the call to ban GM crops as undemocratic: "give [farmers] a choice, is that not what democracy is all about?" Farmers should be able to choose and decide for themselves, he asserted.[28] Food Sovereignty Ghana has made claims about the undesirability of GM seeds for all farmers in the country, framing seeds as an "imposition on Ghanaians" on their website. Although they have the support of a farmers' organization in Brong-Ahafo, as of August 2018 they had not consulted with farmers in northern Ghana.[29] If Food Sovereignty Ghana succeeds in obtaining an injunction to halt GMO commercialization without the participation of farmers in the North, is this really food sovereignty? What is the way forward when considering technologies that may respond to certain needs (high reliance on imports, ageing demographic of farmers, population expansion) but create new risks?

While at the Savannah Agricultural Research Institute, I asked agricultural research scientists about their views on GMOs. During an interview, one of the scientists expressed concern not only regarding GMOs but about the fact that he was the one being asked to assess this controversial technology. "I think the major problem here is when you say GMO it raises ethical and other questions"; he hesitated, and then continued, "whether the scientist may not be the best to answer those questions even though he is at the forefront of the debate."[30] Ghana's biosafety regime places considerable power, however ambivalently held, in the authority of scientists like him, suggestive of how concerns regarding biosafety and biodiversity are shaped by a rule of experts (Mitchell 2002). Determining the safety of genetically modified crops has so far operated through a constriction of vision: a focus on testing in confined field trials for gene flow (what I have described as "genecentrism," as discussed in Chapter 3), to the neglect of the socioeconomic and ethical impacts of the technology. The Ghana Biosafety Act, which governs biotechnology in Ghana, formalizes this genecentric view. I was told by the acting CEO of the National Biosafety Authority that a consideration of socioeconomic concerns is the "duty of the 'anti-group' whatever are the downsides of GMOs."[31] In the view of the authority, such concerns were to be raised by "civil society" and activists after the confined field trials are completed and approved, and before placing the technology on the market. Such a statement reflects an abdication of state responsibility.

Food Sovereignty Ghana, which has raised such concerns, has turned to the same biosafety legislation to demand a ban on the commercialization of GMOs, de facto legitimating the mechanism that allowed GMOs into the country in the first place. This maneuver of bringing the GMO debate to the courtroom reinforces a rule of experts and has potentially exclusionary, and contradictory, effects for the movement of food sovereignty in Ghana.[32] For example, the dangers of having legal battles with agribusiness corporations are demonstrated by *Monsanto v Schmeiser*, discussed in Chapter 3, as well as what one judge described as Monsanto's "incredibly litigious" nature that has led to Monsanto being awarded more than US $23.6 million in seventy-two recorded judgments for patent infringement in the United States alone by 2012 (Center for Food Safety 2013: 6). As early as 2003, Monsanto had a department of seventy-five employees and a $10 million budget dedicated to these kind of legal pursuits; agribusiness giant DuPont has followed suit (ibid.). Resources like these dedicated to advancing global agribusiness interests provoke the question of how well the Ghanaian state or activists would fare in a lawsuit against an agribusiness giant in battles over patent protection, benefit sharing, and culpability in risks associated with genetically modified crops if laws created with those interests in mind are the only recourse to settle such contestations.

The Cornell Alliance for Science has been a central figure in the biotech proponents' response to the food sovereignty movement in Ghana. In a July 30, 2019, post to their website, they used an image of Indian food sovereignty activist scholar Dr. Vandana Shiva and Food Sovereignty Ghana with the headline, "Activists plan to call Vandana Shiva as 'expert' witness in Ghana GMO court case." Contributor Joseph Opoku Gakpo, a graduate of Cornell University's two-month communication fellowship for Ghanaian journalists (Rock 2018: 2–3), interviewed a spokesman for the Ghana National Association of Farmers and Fishermen (GNAFF),[33] who has joined alongside the National Biosafety Authority and the Ministry of Food and Agriculture as a defendant in the Food Sovereignty Ghana suit. GNAFF spokesman John Awuku Dziwornu critiqued the visit of Shiva to Ghana in a workshop that addressed genetically modified seeds and food sovereignty, "If you claim that introducing GMOs amounts to bringing in multinationals to take over seed systems, and you are bringing in foreigners to defend your case … what is the point?" Gakpo highlighted that "Shiva has been involved in Ghana's anti-GMO campaign for some time now. She came to the country in June 2014 to address a public fora on GMOs, urging Ghanaians to reject the technology." In an October 29, 2019, post, they take their argument further by posing the question, "Is it 'unconstitutional' to let activists use Ghana's court system to ban GMOs?" Rock (2018: 189) describes how the Cornell Alliance for Science has not only provided workshops and training to journalists, agricultural research scientists, regulators, and policymakers but has also hosted an online petition in support of the PBB, counting WACCI, CSIR, GNAFF, and the Biotechnology and Nuclear Agriculture Research Institute as cosponsors in "support of agricultural improvement in Ghana." The petition "appealed to readers that intellectual property rights regime would establish financial incentives for plant breeders to continue breeding, and outlined that the law would protect new varieties *in general*"(Rock 2018: 189).

Anti-GM opponents view the introduction of genetically modified crops as a means of indebtedness, debt being a "mechanism of neocolonialism" according to Nkrumah. Food sovereignty activists have protested on the streets of Accra (and beyond) that the introduction of GM seeds impoverishes both farmers with debt and the environment with ecological degradation. Seed sovereignty, the right of farmers to save, exchange, use, and sell their own seed, is violated by control over seed systems through patents and corporate domination of the supply of seeds. Activists argue that these violations are enabling conditions of, rather than exceptions to, the global GM seed regime. In contrast, proponents of "Green Revolution" programs want to encourage the adoption of improved seed in order to address the problem of "genetic erosion": saving local seed is seen as a bad practice because "genetic potency declines" and "seed fertility declines" over time.[34] This again reflects a genecentric view of biological processes, at odds with the idea of seed sovereignty. From the perspective of seed sovereignty, the seed is more than just the store of genetic information, rather a central part of cultural and biological diversity. The ability of farmers to manage their own seed systems, to store, share, and exchange seeds, is viewed as the key to resilience (Alliance for Food Sovereignty in Africa 2014: 7). Many within the food sovereignty movement see genetically modified seed as a vehicle that will destroy these farmer-managed seed systems, in particular through concerns of market domination and cross-pollination.

VI. Conclusion

The conversations described above reveal that Ghana's agrarian future is indeed contested. That is why solutions to agricultural challenges need to be discussed not behind closed doors with donors but rather out in the open—the subject for Ghanaians to debate freely, ask critical questions, and have a stake in shaping their agricultural future. Ghana, due to its deficiencies in agricultural extension and budget for research, is in a position to welcome external support. But such support comes at a cost to appease donors with a "win-win" of advancing both Ghanaian and donor interests. This practice presumes that each actor involved gets something out of it, but it says nothing of the distribution and the scale of such wins. It exhibits the same logic in free trade debates that conflates equality with equity and says nothing of rich world subsidies, economies of scale, and the biases of intellectual property right protection.[35] Nkrumah would likely be quite skeptical of these so-called win-win development interventions.

Despite widespread contestation over Ghana's agrarian development, there is common ground among a wide array of Ghanaians I have spoken with over the years that there is a need for "African solutions to African problems." As professor and founding director of WACCI, Dr. Eric Danquah told me when I discussed the future of the West African Centre for Crop Improvement following AGRA's decision not to renew its funding, "Africa cannot outsource its science. We need to 'decolonize' science education through efforts such as WACCI where talented Ghanaians do not have to leave the country to get the education they deserve."[36] This language of the *need to decolonize*, to be concerned with "covert coercion" in matters concerning food and agriculture in Ghana today, is shared by actors as different as the founder of WACCI to the executive director of CIKOD.

WACCI is indeed an agricultural research institution founded by a Ghanaian researcher and professor concerned with how to increase agricultural productivity in Ghana. The founding director also holds a Cambridge doctorate in genetics and plant breeding, is the first African to win the World Agriculture Prize (in 2018), is an Alliance for a Green Revolution in Africa grant recipient, and has sought relationships with global agribusiness partners such as Bayer and Syngenta for his graduates to find employment. His interests are driven by concerns about agricultural development in Ghana, participating in the advancement of global agricultural research in plant breeding and genetics, and also in graduating master's students who will be the "managerial class" developing a commercial seed sector in Ghana, as discussed in Chapter 5.

This vision of modernized agriculture and investment capital is not necessarily misaligned with Nkrumah's. Nkrumah (1963: 46) argues that one of the biggest challenges is "how to create a skilled labour force and a body of trained technicians in the many fields of modern agriculture, industry, science and economics in the quickest possible time." If WACCI is able to develop this capacity and curriculum in plant breeding without external agenda setting and attentive to the need to maintain crop diversity—as Nkrumah also stressed—is biotechnology necessarily neocolonialist? The earlier discussion of the ecological vector of neocolonialism suggests how neocolonial relationships can be fostered through ecological change ("pest" infestation, monocultures requiring formulaic inputs developed elsewhere, the insecurities of climate change), even if the introduction of biotechnology is advocated by Ghanaian scientists and farmers alike.

Moreover, my interviews in farming communities in the Northern and Upper East Regions suggest that access to commercial seed is of subordinate concern to priorities of improved soil quality, irrigation, mechanization, road improvements, government support, and secure markets. How pro-poor is this plant breeding curriculum, and how much of this research is driven by those external contributors to the curriculum?[37] How is its research oriented? Furthermore, what are "African solutions to African problems" and who gets to decide that? Which Africans lead in addressing these problems? What is the nature of participation for farmers and citizens in addressing their own problems? How well do these research institutions *listen* to the farmers' challenges and their own means to address them? What does it mean to *decolonize* science education?

This chapter allows us to understand how neocolonial relationships are (re)produced, shaping the future of food and agriculture in Ghana. It highlights that there is a commonly shared anxiety about having agricultural policy driven from the outside. Economic, ecological, and legal vectors create openings for the influence of foreign expertise by way of PPPs and dependencies on technology and technical know-how from the West. As with other development interventions, the "new Green Revolution for Africa" agenda hinges on the argument that the state is insufficient to address its country's development challenges without the markets, funding, knowledge, and advice of an outsider entity. These development interventions that bring technology, knowledge, and training to developing countries do respond to gaps in technology access and compensate for the inadequacies of agricultural extension, yet recognition that the gaps are tied to the gutting of public expenditures on agriculture demanded by structural adjustment policies is also needed. A decolonial reading of this "new Green Revolution" agenda reveals how obstacles to more robust agrarian livelihoods are translated into production, technology, and market access problems, rather than problems of distribution, infrastructure, climate change, and, more generally, power.

The food sovereignty movement has an opportunity to highlight these neocolonial relationships, but so long as it does not remain so disconnected from the rural populations it claims to speak for. My critique of the food sovereignty movement comes out of a desire to see the movement in Ghana become more influential, inclusive, and ecologically sensitive. The more connection such organizations can have with those farming, the more the movements will have a better sense of the shifting political ecology of agriculture in Ghana: one that is influenced by rural-to-urban migration, philanthrocapitalist development, invasive species, and the effects of climate change. Moving forward, the Centre for Indigenous Knowledge and Organizational Development's work fostering connections to farmers and supporting intergenerational knowledge exchange should be bolstered in order to better retain traditional agroecological knowledge systems that are known to be resilient, diversified, and far less resource-intensive. As noted, there is a distance between southern-based food sovereignty organizations and farmers in the North, which is why CIKOD's dedicated work on seed sovereignty in the Upper West is so critical. Food sovereignty activist organizations already play an important role in awareness creation about the potential problems GM seeds pose. However, they need to make themselves more cognizant of the challenges that farmers face and not romanticize traditional farming. Rather, they should seek to understand when and why farmers choose to use technologies that they deem harmful. In order to do so, a more sustained dialogue with and inclusion of farmers in such struggles become crucial to the food sovereignty movement.

CONCLUSION

Contesting Africa's New Green Revolution considers how philanthrocapitalism shapes development in Africa today. Through this study of the micro-politics of the "new Green Revolution for Africa" agenda, I show how efforts in northern Ghana to transform the Ghanaian seed sector and develop "pro-poor" biotechnology advance the interests of the private sector in the Global North. Development planners present these interventions as growing the *local* private sector. However, global agribusiness corporations are well situated to take advantage of these legislative shifts,[1] new distribution networks of seeds,[2] and new forms of education and training[3] that promote the professionalization of farming and the commercialization of the seed sector in Ghana. The outcome of these currently unfolding interventions is uncertain and contested. Yet, the possibility that such interventions may exacerbate poverty and power imbalances rather than remediate them warrants concern if we are to take seriously the pro-poor aspirations of this work.

The narrative of an excluded Africa desperate for a "technological revolution" facilitated by foreign expertise in order to realize its entrepreneurial potential and address ecological threats go hand in hand with the push for legislative change that both recognizes Western forms of property rights and facilitates the privatization of seed commons. Humanitarian appeals of global foundations and agribusiness facilitate this access to local resources (germplasm, knowledge, the influence of "lead" farmers) as well as to policymakers and legislators that develop laws that benefit—even if inadvertently—their interests. I argue that the presentation of this work as philanthropy or aid belies its true interests.

Contesting Africa's New Green Revolution contributes to analysis of the "politics of poverty" (Spann 2017; Escobar 1995: 23) of the "new Green Revolution for Africa" agenda (Moseley, Schnurr, and Bezner Kerr 2015; Moseley 2017; Daño 2007; Scoones and Thompson 2011; Thompson 2014). Chapter 1 speaks to this "politics of poverty" by its emphasis on how international development institutions and funders neglect their own roles in disseminating policy advice that undermined farmers' livelihoods. Why do they think that their advice and priorities should have such sway on African policymakers, experts, and farmers today? When will these global players recognize their own legacies that contributed to the "slow violence" (Nixon 2011) of weakened agricultural institutions of research, extension, and plant breeding? Rather than acknowledge mistakes, these neoliberal development institutions and philanthrocapitalist foundations present opportunities for "win-wins" in the "new Green Revolution," supposedly advancing the interests of both African farmers and investors.

Chapters 2 and 3 critique this logic of the win-win, while Chapter 4 exposes it as part of what I call the political economy of hype. The political economy of hype shows that the promise and potential of the poverty-alleviating properties of new technologies are sufficient for capital's expansion, but there is *no actual requirement* that the technology needs to deliver on its promises for the poor. As another example of hype, in Chapter 5 I turn to the mantra of agricultural development in contemporary Ghana: "to make farming in Ghana a business rather than a way of life."[4] This chapter reflects on how the identity construction of the "serious" farmer blames smallholder farmers for their own poverty and conceals how these new Green Revolution interventions, like many before, serve best the interests of elites and global agribusiness. The last chapter argues that "vectors of neocolonialism" enable the deeper penetration of global agribusiness by way of economy, ecology, and law and reproduce this politics of poverty. Next, I link concepts of philanthrocapitalism and biocapitalism to enduring North–South inequalities and consider some emergent forms of resistance to these novel forms of capital accumulation. I conclude with a discussion of how the notion of "mixing" enriches our understanding of the new Green Revolution in Africa as uneven, incomplete, and hybrid forms of capitalist development.

As farmers increase their adoption of "improved seed" and integrate into global value chains to sell their products as part of the new Green Revolution in Africa, corporations participating in public–private partnerships with national agricultural research institutions are able to access local germplasm and knowledge for private gain. Philanthropy enables this access. Furthermore, much of the money that flows from the global foundations that support food security, nutrition, and public health in Africa flows through institutions, businesses, consultancies, aid agencies, and NGOs that are based in the Global North. A 2014 GRAIN report states that *only 5 percent* of the money granted through the Bill & Melinda Gates Foundation (BMGF) goes to African organizations, with half of the funding channeled through intergovernmental institutions (such as United Nations agencies and CGIAR) as well as the Alliance for a Green Revolution in Africa and the African Agricultural Technology Foundation, discussed earlier (GRAIN 2014a: 3). Of the agricultural grants awarded to universities and national research centers, the foundation gave 79 percent to grantees based in the United States and in Europe and only 12 percent to recipients in Africa, despite its commitments to a new Green Revolution for Africa (ibid.). The BMGF called the report "deliberately misleading" but appeared to affirm this pattern of funding: "The central assumption is that only organisations located in Africa can benefit African farmers—and we think that is incorrect" (Kehoe 2014: n.p.).

The fact that so much money continues to be channeled through Global North entities is a frustrating reality that parallels the moneys that *flow out of Africa* in the form of debt servicing for "development" loans (e.g., George 1988). That is to say that this is part of a pattern of *the appearance of inflows of capital and assistance*—as funding for the new Green Revolution for Africa suggests—when, in fact, it is money that is captured by institutions, researchers, entrepreneurs, agribusiness, and NGOs in the United States and Europe. *Who is developing whom* in these agricultural interventions currently unfolding on the African continent? What is at stake in the capture of new Green Revolution resources by northern institutions, private sector, and NGOs is the further decline and reconfiguration of state capacity and the "privatization by NGO" (Harvey 2005: 177) whereby many functions of the state "have been effectively 'outsourced' to NGOs" (Ferguson 2006: 38).

As Barnett's (2011) critique of the "governance of humanitarianism" highlights, these flows sustain unequal power relationships between donor and recipient as well as the "white-savior industrial complex" (Cole 2012) that obscures how philanthropy itself serves to justify inequality. Take, for example, the appeal of UN World Food Program's Josette Sheeran for the leadership of Bill Gates in "ending hunger now" (Sheeran 2011). Philanthrocapitalist *extraordinaire*, Bill Gates, is presented here as a singular and necessary leader that can help to solve the global food crisis. Activist campaigns—hardly celebrating this outsized influence— similarly view Gates as larger than life. One activist meme circulated by food sovereignty networks in 2010 shows Bill Gates looming as a giant above an African cornfield planted with rows of GM maize; Gates and his foundation are regularly referenced in GRAIN, Alliance for Food Sovereignty in Africa, AGRA Watch, and African Centre for Biodiversity reports.

Despite the Bill & Melinda Gates Foundation's attempts to discredit GRAIN's findings, a search through the grant database of the BMGF confirms these funding trends—and in that missed opportunities to build the capacity of African businesses, consultancies, local forms of assistance, and NGOs to solve so-called African problems.[5] The economic power of Gates and his foundation afford him the ability to bring about changes to "development" given his personal wealth and the foundation's US $40 billion endowment that dwarfs the annual budget of the UN World Health Organization at US $4.42 million (GRAIN 2014a). Agricultural projects in Africa funded by the BMGF are wide-reaching—covering agroforestry; biofuels; women's empowerment in agriculture; crop development of yams, cassava, and cowpea; biotechnology; support for universities; as well as soil fertility and fertilizer extension. However, of this, more than half of the BMGF budget went to entities that promote biotechnology and seed sector reform that—whether intentionally or not—create new incentives and conditions for global agribusiness success. In particular, key beneficiaries of this funding for "agricultural transformation" have been AGRA, AATF, BioCassava Plus, CIMMYT, (Monsanto-funded) Donald Danforth Plant Science Center, Syngenta Foundation, International Food Policy Research Institute, Cornell University, Iowa State University, and Michigan State University—with more than US $607.7 million going to the Alliance for a Green Revolution in Africa alone since its inception in 2006.[6]

Activists would say that it is not surprising that the investment portfolio of the Bill & Melinda Gates' Foundation reveals the purchase of 500,000 Monsanto shares, values of which would grow if commercial soy and the cultivation of genetically modified (GM) crops were to expand across the continent given the dominance of Monsanto/Bayer in the transgenic seed market (Vidal 2010; AGRA Watch 2010). Philanthrocapitalists would argue that all promising investment opportunities need to be pursued in order to have the funds to continue their important work in "impact investing." The "value chain" enterprise approach is seen by these players as an appropriate avenue to maximize their impact, as reflected in the BMGF funding of a $10 million project to "develop the soya value chain" alongside agribusiness giant Cargill in Mozambique (Vidal 2010). However, in thinking about the future implications of these value chains, a key question needs to be considered: *value for whom?* Is work as a contract farmer for GM soy or GM cotton value chains really generating the poverty-alleviating potential and food security as promised? Research on these impacts on farmers in Burkina Faso, Paraguay, or India suggests otherwise (e.g., Dowd-Uribe 2014; Elgert 2016; Oliviera and Hecht 2016; Luna 2019; Sridhar 2006; Desmond 2016).

Another central insight of this book is that philanthropy aids in the access and commodification of biological life in the form of seed and that, rather than being a poverty-alleviating strategy, "seed sector transformation" and "value chain integration" are capital accumulating strategies. A 2012 conversation with a DC-based development planner focused on biotechnology sheds light on how American agribusiness corporations thought about these new Green Revolution interventions early on. When asked how industry players were approaching entering into West African markets, he said that the likely first GM products to enter would be the "tried and true Monsanto/Pioneer maize and soy because the technology has been developed already. ... Their considerations would be the business perspective ... and another consideration would be the regulatory environment."[7] When asked about what kinds of varietals were being developed for the West African context, he briefly mentioned the development of "public sector" crops such as transgenic cowpea and rice but pivoted back to industry concerns: "The focus would be on insect-resistance management. ... There's concern about these crops going to West Africa because you need to plant a refuge in order to use these crops ... 5% has to be a refuge of plants that are suboptimal (non-GM). ... Industry is concerned with 'product stewardship.'"[8] This story raises several key points that echo key concerns of the book: that these new Green Revolution interventions are not pro-poor or context-specific; that, rather than a revolution, they are a series of strategies of management and marketing; and that they are not environmentally sound given the widely acknowledged toxic legacy of Monsanto (however obscured through Bayer's acquisition) and its presence in this new Green Revolution.[9]

This emphasis on the business perspective and a corporate-friendly regulatory environment contrasts with how US development planners and agricultural research scientists in Ghana have presented these relationships with global agribusiness as pro-poor—helping them to develop crop varietals designed for the needs of African farmers and African consumers' preferences. Rather, here we learn that the industry plan has been to market products that are already in existence and is only marginally concerned with "public sector" crops. It also suggests additional costs for farmers that are not well highlighted within debates about GM crops on the African continent, such as the added expense of having a refuge zone of crop sacrificed for insect destruction. It reveals that internal industry concerns about new product development in Africa have been about business interests and insect resistance, rather than the development of a kind of catch-all remedy to problems ranging from poverty, drought, and fall armyworm infestation as we see in presentations of the Monsanto-supported Water Efficient Maize for Africa.[10]

As argued earlier, a lot of the same players, logics, and practices involved in the first Green Revolution are advancing this new Green Revolution. However, here I want to reiterate that the intensity and means of efforts to expand biocapital—through philanthropy and regulatory changes—are distinctive. Kloppenburg (2004: 169) reminds us how the onset of exports of commercial seeds into developing countries led to an opening of access to local germplasm extracted to advance private plant breeding: "The pattern of plant genetic transfer between North and South has been largely unidirectional: from the Third World to developed nations." I argue that what we see in Ghana raises concern about whether there are not similar dynamics underway, with legislative changes that are unlikely to alleviate poverty, support capital-poor farmers' market channels, or protect both local plant breeders' intellectual property and seed commons. So, how to confront the privatization of seed commons? How can communities protect against the exploitative dynamics of biocapitalism? And what alternative models of humanitarianism exist that can disrupt the reproduction of unequal power relations inherent to philanthropy?

Global agricultural biotechnology has faced two primary obstacles in its product development and market expansion: global resistance to GMOs and insect resistance. African markets offer global agribusiness areas for expansion *as well as* greater access to unique germplasm for product development. As discussed in Chapter 1, access to plant genetic material from developing countries forms the very basis of "Green Revolution" innovations (Richards 1985: 146; Kloppenburg 2004; Scott 1998). Graddy-Lovelace (2018: 2) reveals that "as climate change becomes more severe and erratic, genetically homogenous agriculture becomes more precarious. Commercial breeders scramble to find new traits" from gene bank repositories, as well as tapping into traditional and local knowledge of the use of plants. Communities that care for these plants need to be recognized and their generations of experimentation, care, and cultivation in situ (in its original place) need to be *supported* rather than *extracted* for private gain (Graddy-Lovelace 2018; Hill 2017).

One way of doing so would be to rethink philanthropy in a way that rejects the logic of extroversion (Hountondji 1997b) and draws upon local and indigenous practices of reciprocity and lending. *Ubuntu* is one such example: "*ubuntu* stresses the importance of community, solidarity, caring, and sharing. This worldview advocates a profound sense of interdependence and emphasizes that our human potential can only be realized in partnership with others" (Ngcoya 2009: 1). The edited collection, *Philanthropy in South Africa: Horizontality, Ubuntu and Social*

Justice (Mottiar and Ngcoya 2018), suggests that *ubuntu* provides insights to challenge dominant trends in philanthropy in Africa today, considering rather "the beneficent spirit of multitudes of Africans whose acts of generosity sustain millions of their compatriots." What would philanthropy look like if guided by social justice and interdependence? What if the public–private partnerships were *ubuntu*-inspired and endogenous to African societies, supporting local institutions, businesses, experts, intellectuals, consultancies, and NGOs rather than their Euro-American counterparts?

As new Green Revolution discourses proliferate, so do counter-discourses: not only *ubuntu*-based philanthropy but also bio-community protocols and endogenous development. Community protocols in Ghana, such as the CIKOD-supported Tanchara Bio-Cultural Community Protocol, protect sacred groves—sites of bio-cultural diversity that have spiritual importance to the Tanchara community—from the pressures of agricultural and housing development expansion. In a similar fashion, seed sovereignty campaigns in Ghana, such as those that CIKOD has organized, draw from the Nagoya Protocol on Access to Genetic Resources and the Fair and Equitable Sharing of Benefits Arising from their Utilization (ABS), the first binding protocol[11] that "establishes a set of rights of indigenous peoples and local communities ... over their genetic resources and traditional knowledge" (Natural Justice and ABS Capacity Development Initiative n.d.: 7). These initiatives work to secure recognition, prior informed consent, and benefit sharing for any products derived from local germplasm and knowledge—or even the repatriation of indigenous seed (Mgbeoji 2006; Shiva 2001; Graddy-Lovelace 2020; Nagoya Protocol 2010).

The recognition and protection of the value of indigenous and local communities' networks of reciprocity and care, knowledge and cultivation practices, and community ecological governance are found in programs in northern Ghana, such as the intergenerational knowledge exchange programs of RAINS and CIKOD that connect youth with elders to learn traditional agroecological practices, the study of integrated community development at the University for Development Studies, as well as the study and practice of ancestral and traditional knowledge of the use of plants at the Millar Institute for Transdisciplinary and Development Studies and the Taimako Plant Center, respectively. Such programs, curricula, and practices reinforce the importance of agroecological or permaculture principles and reflect forms of endogenous development. They resist the agricultural de-skilling tendencies of monocultural practices that treat farmers as *users* rather than *innovators* or *coproducers* of agricultural advancements. Such reversal in knowledge production trends—where "people's science"[12] (Richards 1985) is practiced and agricultural skills can be developed in service to strengthening the ecosystem and regenerative care—is encouraging and indicates that the transition "from farming as a way of life to a business" will not be totalizing.

As argued throughout the book, it is unlikely that the effects of the new Green Revolution will be profoundly transformative, in contrast to claims made by both its proponents and opponents. Rather, the adoption of these new technologies and practices will face the diverse concerns of African farmers that prioritize feeding their families first over production for export, draw from traditional knowledge and ancestral wisdom to know when and what to plant, and diversify their livelihood strategies to reduce risk. This is not the only mixing relevant to our analysis: philanthrocapitalism and biocapitalism, too, are forms of mixing: the mixing of philanthropy with the logics of venture capital, the "traffic" between the biological and economic spheres (Cooper 2008). However, an important distinction between these forms of mixing needs to be made—while one form of mixing diversifies and mitigates risk, another commodifies and accumulates capital. While the former enacts the agrarian economy as sets of heterogeneous practices and knowledges (e.g., Escobar 2008: 108), the latter seeks to negate its own internal diversity in order to conceal the incompleteness and unevenness of capitalist development. Recognition of this heterogeneity and mixing is critical for those activists and development planners alike that, for different reasons, tend to downplay the independence and pragmatism of African farmers.

NOTES

Preface

1 Interview, May 4, 2012.
2 The oil from this plant is used for biofuel.
3 Mr. Nyari is also affiliated with the African Biodiversity Network.
4 Interview with Nyari, January 10, 2012.
5 Interview, July 5, 2017.
6 Interview by author and Cecelia Lynch, September 2, 2012.
7 This is in reference to former UN secretary-general and founding chair of the Alliance for a Green Revolution in Africa (AGRA), Kofi Annan, promoting the newly formed alliance as a way to "bring about a uniquely African Green Revolution that will unleash the continent's agricultural potential," discussed in Chapter 2 (AGRA n.d.e).
8 Philanthrocapitalism is the merging of philanthropy with the logics of venture capital.
9 One example of this distributed agency is the attention paid to appeasing the ancestors through the offering of libations for a good harvest. See Ignatov (2017).

Introduction

1 Informants involved in mobilizing for and against GMOs confirmed this account that there was little public awareness within Ghana at the time of the passage of the Biosafety Act 831 (interview with lead anti-GMO activist, February 14, 2014; interview with official from Program for Biosafety Systems, May 5, 2015). Friends of the Earth did attempt to raise awareness of the issue prior to the passage but was unsuccessful at gaining widespread attention to the GMO issue in Ghana.
2 This Act allows private seed companies to produce foundation seed, previously solely a public function of the national agricultural research institutions.
3 Interview with consultant for the Ghana Seed Task Force, August 8, 2018.
4 The US Office of Technology Assessment considers biotechnology to include any technique that uses living organisms to improve plants, to make or modify products, or to develop microorganisms for particular uses (US Congress, Office of Technology Assessment 1988: 3).
5 Discussed further in Chapter 6.
6 Interview with GNAFF spokesman, May 15, 2015.
7 Interview with CPP spokesman, May 4, 2015.
8 Interview with food sovereignty activists, May 5, 2015.
9 Formerly African Centre for Biosafety.
10 Throughout the book I clarify both the actors that circulate such discourse and how it is contested.
11 This is seen in the development of drought-, water-, and salinity-tolerant crop traits. However, the effects of climate change may present all of these conditions possibly in

one growing season (periods of heavy rain followed by flooding, for example). A crop "engineered" to be drought-tolerant would suffer under conditions of heavy rain.

12 This is also reflected in the increase in magazines focusing on investment in Africa: *Fortune Africa, African Business, African Banker*. Another example of this excitement over African investment is seen in the BBC award-winning documentary *Africa: Open for Business* (2005).

13 From 2006 to 2010 Ghana's foreign direct investment inflows increased from US $636 million to US $2.53 billion at a compound annual rate of 41 percent. Ghana "has been recognized as one of the most open economies in sub-Saharan Africa for foreign equity investment" (USAID 2012). Tech sector giants such as the CEOs of Twitter and Square, Facebook, Google, and Microsoft are all turning their attention toward investing in African tech start-ups; the CEO of Twitter and Square, Jack Dorsey, says, "Africa will define the future." Microsoft, Facebook, and Google are all involved in the continent with what's known as "accelerator programs"; Visa, MasterCard, and Salesforce are making venture investments in African start-ups. Tech start-ups that utilize GPS and weather forecasting are booming areas in new ICT apps for agriculture (Rooney 2019).

14 Monsanto was acquired by Bayer in June 2018. I refer to Monsanto because that is the corporation that "donated" the transgene for the development of *Bt* cowpea at the Savannah Agricultural Research Institute.

15 I am grateful to Sara Berry for this point.

16 This became apparent after multiple conversations at SARI regarding *Bt* cowpea—it was unclear which cowpea gene would accept the transgene and such results were found after trial and error. The difference is that biotechnology can speed up certain processes through the identification of marker genes. On farmer experimentation and "people's science," see Richards (1985).

17 This is not to say that local control is not subject to elite capture, corruption, or gender/minority-based discrimination.

18 Most directly addressed in Chapter 3.

19 The Food and Agriculture Organization of the United Nations (FAO) identifies four pillars of "good agricultural practices": economic viability, environmental sustainability, social acceptability, and food safety and quality (FAO n.d.b). In practice, most GAPs are de facto determined by Western countries.

20 Whereby legislation in countries such as Ghana align with the requirements under the TRIPS Agreement toward a system like the International Union for the Protection of New Varieties of Plants (UPOV) Convention or a similar *sui generis* system that protects intellectual property (Manu 2016).

21 This concern was institutionalized in 2010 through the Principles for Responsible Agricultural Investment sponsored by the World Bank, the FAO, and the United Nations Conference on Trade and Development (Clapp 2014).

22 The reproductive material of a plant, in the form of seed or plant tissue, collected for research, breeding, and conservation.

23 A locally adapted, domesticated variety.

24 Harriett Friedmann coined the term "food regime," defining it as "the rule-governed structure of production and consumption of food on a world scale" that emerged in the post-war period (Friedmann 1993: 30–1).

25 "*Homo œconomicus* is an entrepreneur, an entrepreneur of himself" (Foucault 2008: 226).

26 A key tenet of food sovereignty.

27 Another analogous example of this lack of agency is reflected in agricultural interventions that focus on the "last mile user," discussed in Chapter 5, whereby farmers are reduced to "users" rather than generators of technology.

28 Interpretivism is best suited for studying meaning-making processes and how social and political identities crystallize or change over time. For more on the interpretive turn in political science, see Yanow and Schwartz-Shea (2006) and Wedeen (2010).

29 Foucault (1972: 49) considers discourse "as practices that systematically form the objects of which they speak."

30 This idea of "deficiency" within development discourse refers to a lack of income, technology, market access, or education. Deficiency is conceived of as a problem of the poor catching up with the rich; deficiency is in this sense a relative concept. See Sachs (2010b).

31 Some of the texts analyzed include annual reports of foundations and agribusiness corporations; the policy plans of bilateral aid agencies, intergovernmental organizations, or NGOs; congressional hearings on plant biotechnology in Africa; interview transcripts; public statements; and websites featuring advertisement and political campaigns by seed corporations and activist organizations. I examine key texts that have been enlisted by major actors in the GMO debate (which I have been studying since 2004) or in development planning that are anticipated to reach a wide audience (such as World Bank reports, protest statements by social movement organizations, or annual reports of foundations). To study the debate over the cultivation of GM crops in Africa, I tapped into online advocacy networks via social media such as Greenpeace, La Via Campesina, African Centre for Biosafety, Food Sovereignty Ghana, Slow Food International; tracked new developments in the agricultural biotech industry via the listservs of the ISAAA, Monsanto, World Poultry, All About Feed; established a Google Alert of "GMO," "biotech outreach," and the "African food crisis"; conducted LexisNexis searches on food security and biotechnology; monitored the Convention on Biological Diversity Biosafety Clearinghouse; followed Ghanaian news sources such as Ghana News Agency, *Daily Graphic*, All Ghana News, and My Joy Online; and analyzed the reports of major actors in agricultural development such as the World Bank, USAID, and USDA GRAIN, annual reports of Bill & Melinda Gates and Rockefeller Foundations, Feed the Future reports, and the New Alliance for Food Security and Nutrition progress reports.

32 Totaling twelve months. 2012–13 is when I completed my Fulbright fellowship. My follow-up trip in 2015 was funded by the UMD Program for Society and the Environment; funding for travel in 2018 was provided by Appalachian State University. Pseudonyms are used throughout where fieldwork is used.

33 The AU Maputo Declaration makes it evidently clear that the major cutbacks to agricultural spending were misguided, committing African states to dedicate 10 percent of their budget to agriculture. AGRA has been working with African governments and development planners to promote "smart subsidies" (that stands in contrast to structural adjustment programs that cut them). In 2016, the IMF recognized that living with debt can be preferable to austerity (Ostry, Loungani, and Furceri 2016: 40).

34 Whether or not it will be enduring is another question as many prior efforts to promote regular purchase of "improved" seed and fertilizer have not often lasted beyond the period of subsidy benefit. See Chapter 1.

35 Interview with development expert involved in the curriculum development, August 15, 2018.
36 Although climate's impact on agriculture is globally felt.

Chapter 1

1 For further discussion of the origins of the term, see Patel (2013).
2 For more on legibility, see the introduction and Scott (1998).
3 It is important to recognize that other (non-high-yielding) crops also experienced productivity gains (Patel 2013; Paddock 1970). Paddock (1970: 897) shows that between 1967 and 1970, Indian production of barley, chickpeas, cotton, tea, tobacco, and jute also increased by 20–30 percent.
4 Berry (2002) explicates that the European colonial rule impacted land access in three primary ways: through displacement of Africans, through attempts to draw and redraw territorial boundaries, and through efforts to codify custom and customary land access regimes (i.e., the "invention of tradition" that constructed land as communal property of tribes and empowered chiefs as its administrators). The colonial legacy on land rights, in short, created divisions between the landed and the landless (Cooper 2002: 23), enhanced the rent-seeking capabilities of traditional rulers (e.g., Mamdani 1996), and intensified claims over what custom is (Berry 2002). Late colonial efforts to modernize African agriculture assigned land to "forward-looking" farmers and created new divisions between "progressive" and "traditional" farmers (Berry 2002: 647). These divisions created the conditions of possibility for the processes of social differentiation that I explore in Chapter 5. Since independence, Ghana has undertaken no profound land tenure reform (Amanor 2008: 68).
5 This is in reference to the literature on appropriate technology (e.g., Schumacher [1973] 2000).
6 Joint Sector Review meeting in Accra, June 30, 2017.
7 Presentation by Northern Region regional director of Ministry of Food and Agriculture Plant Protection Unit at the Mango Value Chain workshop, Tamale, Ghana, July 18, 2012.
8 For example, the UN Millennium Development Goals generated some of the political will to reverse policies that cut social spending.
9 Interview with Agriculture Technology Transfer Project official, August 8, 2018.
10 This is in reference to former UN secretary-general and founding chair of the Alliance for a Green Revolution in Africa (AGRA n.d.e), Kofi Annan, promoting the newly formed alliance as a way to "bring about a uniquely African Green Revolution that will unleash the continent's agricultural potential," discussed in Chapter 2 (Annan 2006, available www.agra.org/who-we-are/our-history/, accessed June 6, 2014).
11 When this was initiated, it was G8 countries; Russia is no longer a part.
12 McKeon (2014) puts this number at over hundred companies if the agribusinesses involved in New Vision and Grow Africa are included.
13 In a similar vein, indigenous knowledge of the Chhattisgarh peoples was used for work on high-yield varieties of rice in India at the Madhya Pradesh Rice Research Institute (MPRRI). According to Shiva (1991: 44), the MPRRI was "closed down due to pressure from the World Bank (which was linked to IRRI through CGIAR) because MPRRI had reservations about sending its collection of germplasm to IRRI."

14 As Kloppenburg (2004: 172) shows, "Delegates from Third World and industrialized socialist countries called for the application of the principles of common heritage and free exchange to *all* categories of germplasm" (emphasis in original).

15 Interview with Program for Biosafety Systems official, May 5, 2015.

16 Interview with US State Department foreign service officer, March 22, 2013.

17 The same can be said of the discourse of "good governance." Mkandawire (2007: 680, 681) discusses the African origins of this term (which was understood within African intellectual circles to refer to state–society relations that are developmental, democratic, and socially inclusive) that have now been emptied of meaning by development planners whose use refers to "very much business as usual."

18 Such a search (no date restrictions) garnered 11,705 results and included results where "enable" and "environment" were used.

19 Original text was both bolded and italicized.

20 Emphasis in original.

21 Interviews with USAID official, June 22 2017; interview with GCAP official, July 30, 2018.

22 Interviews with Agriculture Technology Transfer Project official, July 27, 2018; USDA officials, August 6, 2018; and US State Department biotechnology official, March 22, 2013.

Chapter 2

1 The "Presentation on Sector Performance: Agricultural Sector Progress Report, 2016" at the Joint Sector Review meeting in Accra on June 29, 2017, indicated that in 2016 the released funds from the annual budget funding amount from the Ministry of Finance was only 52.6 percent of budgeted funds for the Ministry of Food and Agriculture; internally generated funds released only 60.4 percent. By contrast, donor funds released 102.5 percent of budgeted funds in 2016.

2 Interview with AGRA official, August 2, 2018.

3 This ratio is widely recognized as inadequate and ranges between 1:1,000 and 1:3,000, with 1:1,500 being the most commonly cited statistics in the interviews I conducted.

4 Joint Sector Review meeting, June 29, 2017.

5 Interviews with women cowpea farmers in the Northern Region, July 5, 2017; interview with agricultural research scientist at the Savannah Agricultural Research Institute, July 19, 2017. Zerbe (2005) makes a similar point about the lack of attention to women's views on crop traits in the development of biotech cassava.

6 Interviews with faculty at WACCI, August 3 and 6, 2018.

7 TechnoServe has received $85 million from the Bill & Melinda Gates Foundation and works in partnership with multinational corporations such as Nestle, Coca-Cola, Cargill, and Unilever (GRAIN 2014a: 9).

8 Muraguri is referencing one of the first agricultural biotechnology partnerships between KARI and Monsanto in developing a virus-resistant sweet potato. As a result of a disconnect between the farmers and the product developers, the sweet potato was engineered to be resistant to a virus that was uncommon in the area and did not have the intended effect of reducing crop loss.

9 The first commercial soyfood was a soy-enriched wheat bread in 1962, part of Adventist relief work in Ghana. US PL 480 distributed soy-based food aid to Ghana in 1971, 1973–9, and 1982 (Shurtleff and Aoyagi 2014).

10 I consider the International Fertilizer Development Center a philanthrocapitalist organization because its anti-poverty work is to get farmers to increase their fertilizer uptake, with obvious benefits for the fertilizer industry (https://ifdc.org/partnerships/, accessed December 21, 2020).

11 Interview with ATT official, July 27, 2018.

12 Interview with a lead researcher for the SIL, August 15, 2018.

13 The key lobbying arm of the soybean industry in the United States.

14 Interview with USDA official, June 28, 2017.

15 The dominant soybean production model in the United States uses genetically modified seed almost exclusively, and so this promotion of soy could create another entry point for GMOs in Ghana.

16 Monsanto acquired Climate Corps and as of June 19, 2019, was actively hiring in data analytics, engineering, software development, and web management positions in the San Francisco Bay Area.

17 I am grateful to an anonymous reviewer for help with this articulation.

Chapter 3

An earlier version of this chapter's argument has been published. See J. A. Ignatova (2017), "The 'philanthropic' gene: Biocapital and the new Green Revolution in Africa," *Third World Quarterly*, 38(10): 2258–75.

1 By new market value, I refer to the changing valuation of seed that facilitates the formation of agrarian capital. Drawing upon authors such as Birch (2017), Kloppenburg (2004), Polanyi ([1944] 2001), and Sunder Rajan (2006), I understand value as a social practice. Likewise, I view property as a social process of negotiation of access to material (or intellectual) resources (e.g., Berry 1993).

2 Cowpea is both versatile and significant to northern Ghanaian diets—a legume that can be prepared and eaten as either an ordinary dish (*waakye*, rice and beans) or as a special treat (*tubaani* eaten during holidays or Sundays).

3 Interviews in Washington, DC, May 4, 2012, and at SARI in Ghana, May 7, 2015.

4 Brooks (2005: 367–8) makes a similar argument.

5 The creation of biocapital can also be seen with research on the genome as well (e.g., Smith's (2012a) work on rice genomes). I have focused on genes because my informants involved in agricultural research underscored a concern about the gene (the proprietary *Bt* transgene), with a general tendency toward a reductionist (rather than a systems-thinking) approach.

6 Interview with National Biosafety Authority official, May 20, 2015.

7 The majority ruling rejected the comparison to the *Harvard College v Canada (Commissioner of Patents)* Canadian Supreme Court decision on the "OncoMouse," which determined that higher order beings could not be considered merely compositions of matter patentable under the Canadian Patent Act.

8 However, much of the product development occurs upstream in research institutions like CSIRO Australia or by companies like Monsanto. Interview with Program for Biosafety Systems official, Accra, May 5, 2015.

9 For more information about other philanthrocapitalist agricultural biotechnology PPPs across the African continent, see Ignatova (2015: 183–6).

10 Interview with lead agricultural research scientist on cowpea at SARI, February 13, 2013.

11 Ibid.

12 Interview with official for PBS, May 5, 2015.

13 Interviews with agricultural research scientists at SARI, February 13, 2013; February 22, 2013; and May 7, 2015.

14 E-mail correspondence with AATF, October 27, 2014.

15 Interview with senior advisor to PBS, May 15, 2015.

16 The New Alliance identified implementing new policies related to seed and fertilizer use, as well as appointment of private sector representatives in grain value chains, as key policy commitments of the Ghanaian government to improve agricultural productivity (G8 New Alliance for Food Security and Nutrition n.d.).

17 Interview with a senior advisor to PBS, May 15, 2015.

18 The PBB has since been retitled the "Plant Variety Protection Bill"; however, it raises the same concerns discussed here and in Chapter 6, as it has not been substantially revised. Ghanaian parliament passed the bill on October 9, 2020. As of December 21, 2020, the bill is awaiting presidential assent (Parliament of Ghana 2020).

19 Nigeria drafted a Plant Variety Protection Act that extends similar intellectual property rights to plant breeders in March 2019 (AGRA 2019). Interview with a senior advisor to PBS, May 15, 2015.

20 Interview with biotechnology advisor for USAID in Washington, DC, May 4, 2012.

21 When the Rockefeller Foundation attempted to bring herbicide-resistant maize into Kenya, "negotiations broke down after Monsanto demanded full ownership of all future research results derived within Kenya, a demand that Kenyans understandably refused" (Paarlberg 2001: 48).

Chapter 4

1 For example, a LexisNexis search of "food" and "crisis" between 2007 and 2011 yielded over one thousand relevant hits, most of which identified the role of the 2007–8 food price crisis in creating food insecurity.

2 Emphasis in original.

3 Emphasis in original.

4 My emphasis.

5 Genetic use restriction technology was initially developed by a subsidiary of Monsanto, a major producer of genetically modified seeds, together with the USDA, in order to prevent farmers from saving seeds for subsequent harvests. Following activist pressure, Monsanto pledged in 1999 not to commercially release this technology (Kloppenburg 2004: 10–11). Commercial varieties of GM crops pose quite a different problem—that of the cross-pollination of transgenic varieties with native varieties of the same crop, what is known as "biopollution," and was a central issue in the 2004 Canadian Supreme Court case *Monsanto Canada Inc. v Schmeiser*. The mechanism that prevents the saving of GM seeds is through patents and "technology use agreements," contracts that forbid farmers who purchase their seeds to save them for future use.

6 The use of the term "myth" to disparage the work of the opposition is used by both proponents and opponents. GM Watch lists "myth-makers"—those individual scientists and consultants as well as organizations—that promote the myth that GM foods are safe and desirable. In contrast, Monsanto also has a dedicated page on its website "Myths about Monsanto" where it addresses and "corrects" these myths.

7 Interview with agricultural research scientists at SARI, February 13, 2013.

8 Interview with biotechnology outreach official at SARI, February 13, 2013; interview with official from Forum for Agricultural Research in Africa, March 6, 2013.

9 Director general of the CSIR Ghana keynote address at the one-day "sensitization seminar of agricultural biotechnology and the *Bt* cowpea project," September 18, 2013, at SARI, Nyankpala, Ghana. Transcript provided by an agricultural research scientist at SARI. There is no indication that these groups are "heavily funded."

10 Interview with agricultural research scientist at SARI, September 30, 2013.

11 Discussed in the preface. Interview with biotechnology advisor at USAID, Washington, DC, May 4, 2012.

Chapter 5

1 Interview with a Northern Region district assemblyman, September 18, 2012.

2 By "elite" I refer to not only capital-rich farmers with political connections but also farmers who have a history of connecting with these agricultural development interventions by NGOs. My preliminary observations indicate that elite farmers stand to gain more from access to "improved" and genetically modified seed, finance, and legal reform.

3 My emphasis.

4 I use the term "traditional" as many chiefs and elders I spoke to in my research in farming villages identified as "traditionalists;" I treat this as synonymous to peasant or subsistence farmers, self-provisioning farmers, smallholder farmers (though the latter has been misused by agencies like the World Bank to include farmers with up to 100 ha of land, certainly not the "smallholders" I encountered in northern Ghana).

5 This is a reference to "sensitization" seminars that I discuss in Chapters 4 and 6.

6 Interview with cowpea farmers of Cheyohi, July 5, 2017; interview with Tamale seed producer, July 20, 2017.

7 Interview with Bakari Sadiq Nyari, February 13, 2013. The Gurene language of the Upper East Region names the days of the week according to their sequence in relation to market days. This linguistic organization of time reflects markets' importance and the fact that people do generate enough surplus produce to trade in markets, just not in the markets of the global food economy. The "market" in the context of the term "non-market system" is akin to what Polanyi ([1944] 2001) describes as the "market economy."

8 Interview with opinion leader in Kukuo Yapalsi, September 18, 2012.

9 Interview with official from the Centre for Indigenous Knowledge and Organizational Development, August 4, 2017.

10 As discussed in Chapter 1, efforts to change agrarian practices in northern Ghana have a long history that include colonial agricultural interventions that promoted the expansion of plantation economies as well as state-led expansion of large-estate farming in northern Ghana.

11 Interview with USAID official, June 22, 2017; interview with GCAP official, July 30, 2018.

12 Interview with farmers in Nyariga, July 24, 2018; interview with IFDC official, August 8, 2018.

13 Interview with Ghanaian climatologist, March 10, 2013.

14 Interview with Upper East regional director of the Ministry of Food and Agriculture, July 4, 2017.

15 However better tailored to specific crops and agroecological zones through support from AGRA's Soil Health program.

16 Discussed in the introduction (FAO n.d.b).

17 Interviews with ADVANCE officials, February 18, 2013, and July 26, 2017.

18 Interview with director of WACCI, August 6, 2018.

19 The University of Illinois, Urbana-Champaign, is also the site of the Monsanto Innovation Center. "Monsanto is also a longtime partner with Iowa State University" (https://www.foundation.iastate.edu/, accessed December 20, 2020).

20 Interview with development expert involved in the curriculum development, August 15, 2018.

21 The cost estimate of the four-year program is an astounding US $118,898, with four years' tuition and registration alone accounting for US $13,699, making it well outside the reach of most Ghanaians without external support (http://wacci.ug.edu.gh/content/wacci-phd-fees-schedule, accessed December 20, 2020). This also suggests that the funding would come from an international institution rather than a domestic one.

22 Although nGR initiatives place a lot of emphasis on gender inclusivity, men are still overrepresented in commercial farming in northern Ghana.

23 Term refers to the process of installing a chief in northern Ghana, in reference to the animal skins upon which chiefs sit.

24 Interview with ADVANCE official, July 26, 2017.

25 For a discussion of environmental and social learning in the case of genetically modified cotton seeds, see Stone (2007).

26 The interlude expands on this point.

27 Interview with chief of Vea, December 11, 2012.

28 Interview with ADVANCE official, February 18, 2013.

29 Interview with young large-scale rice farmer, Northern Region, May 9, 2015.

30 This was directly contradicted by a district Ministry of Food and Agriculture director in northern Ghana, who said, "The seeds are there, the seeds are not a problem. Agricultural machinery is a problem" (interview, March 11, 2013).

31 It is also derivative of the language of "last mile delivery" and "Project Last Mile" used to describe the reach of companies such as Amazon and Coca-Cola's public–private partnership with USAID to deliver vaccines, respectively.

32 Interview with ATT Project official, July 27, 2018.

33 Regarding the interventions "to build research capacities and promote labor saving technologies," my observations indicate that the latter was more successful than the former.

34 Interview with ATT official, July 27, 2018 (Feed the Future Ghana n.d.).

35 Ibid.

36 Ibid.

37 With Digital Green and Access Agriculture.

38 Interviews with ATT official, July 27, 2018; ADVANCE official, July 26, 2017.

39 Interview with development expert promoting commercial seed sector in northern Ghana, August 15, 2018.

40 Interview with ADVANCE official, July 26, 2017.

41 Interview with principal investigator of SIL, August 15, 2018.

42 Interview with development expert involved in the curriculum development, August 15, 2018; also confirmed in interview with ATT official, August 8, 2018.

43 Ibid.

44 Interview with IFDC official, August 8, 2018. This meant that USAID and AGRA had successfully gotten the Plant Protection & Regulatory Services Directorate (PPRSD), which controls seed inspection and certification, to agree to allow private seed inspection through their offer of seed labs and moisture meters that would assist the PPRSD with inspection, ultimately moving toward "seed inspection on a for-profit basis" (ibid.).

45 Subject to registration and certification with PPRSD.

46 Interview with ADVANCE official, July 26, 2017.

47 Time will tell whether farmers will continue to purchase seed and supporting inputs as many communities only recently received the free starter packages.

48 Interview with ADVANCE official, July 26, 2017. It should be noted that ADVANCE does not work with subsistence farmers.

49 Ibid.

50 The language of "commercial farmer" is more commonly used than "nucleus farmer" in the GCAP scheme, but connotes the same meaning, and in other contexts has been used interchangeably.

51 Interview with GCAP official, July 30, 2018.

52 Ibid.

53 Interview with GCAP official, August 3, 2017. Although, at its onset, they invited select northern Ghanaian farmers to participate.

54 The Vea Irrigation Project was constructed in 1965 and has had no major rehabilitation since.

55 On numerous occasions, development planners were surprised when I shared that local people had attempted to address the dam's problems by themselves—they had just assumed that these communities were waiting for the government for help.

56 Interview, July 24, 2018.

57 Interview with Lands Commission official, October 28, 2013.

58 This is akin to what Chambers (1983: 18) identifies as elite bias.

59 Interview, July 27, 2018.

60 This would include transgenic cowpea, if it is approved for commercialization (interview with principal investigator of *Bt* cowpea project, SARI, May 7, 2015).

61 When I met him at Heritage Seed, he was incredibly welcoming and introduced me to his wife as a partner. He talked about the ways in which he utilizes traditional methods of farming to reduce the use of agro-chemicals, an example of a "mixer" that I describe in the following interlude.

62 Interview with a representative of ADVANCE, Tamale, February 18, 2013.

63 Interviews with agricultural research scientists at SARI, February 13 and 22, 2013.

64 Interviews with USDA officials, June 28, 2017.

65 A number of my agronomist contacts that had traveled to the United States were not favorably impressed by the farming practices in the United States. One remarked that he was struck by how little agricultural diversity was present in the Midwest. Other farmers and biosafety regulators commented in their travels to the American Midwest at how unhealthy Americans appeared and questioned the American diet; one informant characterized the American diet as "sugar-sugar."

66 My emphasis.

67 The World Bank states in a project appraisal document for the Ghana Commercial Agriculture Project that these farmers are not included in plans for commercial agricultural activities such as outgrower schemes: "Small holder, family farms can be commercial if they interact sufficiently with the market (for inputs and especially outputs). Agri-business and agro-processing—large- and small-scale—is also included. It would not include extremely poor marginalized households dependent on subsistence farming under extremely fragile and disadvantaged circumstances. *The opportunities created by this project, for instance participation in out-grower schemes, are unlikely to be accessible because of severe capacity and behavioral constraints.*" The World Bank, Project Appraisal Document on a Proposed Credit in the Amount

of SDR 64.5 Million (US $100 Million Equivalent) to the Republic of Ghana for a Commercial Agriculture Project, February 27, 2012, p. 6 (my emphasis).

68 Interview with Lands Commission official, October 28, 2013.

69 Ibid.

70 "One might say that the ancient right to *take* life or *let* live was replaced by a power to *foster* life or *disallow* it to the point of death" (Foucault 1990: 138). Emphasis in original.

71 Interview with US State Department official, April 9, 2015.

72 Outgrower schemes address climate change through crop varietals that are early maturing and drought and salinity tolerant as well as virtual delivery of rain forecasting through companies such as Ignitia, but they do not address losses from climate destruction.

73 This became apparent when I guest lectured at UDS Nyankpala and met a master's student who was a part of a nucleus-outgrower scheme to produce sorghum for Guinness Ghana. We had talked about the various opportunities and risks of the expansion of commercial agriculture in the North, and the interest rates associated with the loans she was encouraged to take on as a nucleus farmer. Despite her level of education and her evident intelligence, she was shocked to learn that interest rates for agricultural loans, on average, exceeded 30 percent; she had taken out similar loans with little expectation that they would be so high.

Interlude

1 For more detail about Ignatov and Azaare's collaboration, see the feature on Azaare's work in *Africa's* (2020), *Local Intellectuals Series*, 90(4): 649–82.

2 Interview with ATT official, July 27, 2018; interview with USAID official, June 22, 2017; interview with Seed Sector Task Force consultant, August 8, 2018; also reflected in discussions at the 2017 Joint Sector Review meetings in Accra.

3 Interview, March 12, 2013.

4 One of her brothers runs a high school adjacent to the family compound.

5 The social obligations associated with wealth were communicated when my relatively wealthy friends would comment that they were always expected to bring gifts or the like when visiting their extended family or community. Fatimata's embeddedness within her network of social and environmental relationships contrasts with the "market society" that Polanyi ([1944] 2001) describes, whereby the economy is *disembedded* from social and ecological relations.

6 Interview, September 13, 2012.

7 Interview, July 15, 2017.

8 This distinction is further explored in the next chapter.

9 Interview, June 28, 2017.

Chapter 6

1 Allain (2005) defines a vector as "a quantity with more than one element (more than one piece of information)." This also lends itself to the indeterminacy I wish to suggest.

2 Presentation by agricultural director of PPRSD at the Joint Sector Review meeting, June 30, 2017.

3 Some agricultural research scientists, although generally more skeptical of these locally made concoctions, conceded to me that this type of trial and error was generally more benign than chemical cocktails that can pollute the local water and food supply.

4 Presentation by agricultural director of PPRSD at the Joint Sector Review meeting, June 30, 2017; interview with SARI official, July 4, 2017; interview with ADVANCE official, July 26, 2017.

5 Presentation by agricultural director of PPRSD at the Joint Sector Review meeting, June 30, 2017.

6 Interview with official from ADVANCE, July 26, 2017.

7 The Syngenta Foundation offers a $134,000 scholarship to support a vegetable breeder attendance at WACCI's plant breeder doctoral program (https://www.syngentafoundation.org/press-release/recent-news/scholarship-supports-future-vegetable-breeder, accessed December 21, 2020).

8 These university partnerships are discussed in Chapter 5. As a part of the master's program in Seed Science and Technology, students from the University of Ghana participate in a six-week internship in the United States with researchers at the University of Illinois in the Department of Crop Sciences and with private sector companies, including Dow AgroSciences and Monsanto (http://soybeaninnovationlab.illinois.edu/internship-provides-new-opportunities-next-generation-african-plant-breeders, accessed December 21, 2020).

9 It is important to note that the ATT Project does support the development of local agro-dealers; however, the products they sell are where they have opportunities for global agribusiness market expansion.

10 The bill has since been retitled the "Plant Variety Protection Bill"; however, it raises the same concerns discussed in this section and in Chapter 3 as it has not been substantially revised. Ghanaian parliament passed the bill on October 9, 2020. As of December 21, 2020, the bill is awaiting presidential assent (Parliament of Ghana 2020).

11 Interview, August 2, 2017.

12 Although this arguably offers an opportunity for both traditional and indigenous knowledge to gain recognition, it also reflects genecentrism in that the entire treaty is premised around plant genetic resources with those inherent limitations and predispositions I discuss in Chapter 3. Whereas the 68-page treaty has 109 references to "plant genetic resources" or "genetic resources," "traditional knowledge" has 1 (within Article 9 on Farmers' Rights), "indigenous" has 2 (Article 5.1 and Article 9.1), and "resources" has 123. (Twenty-four references to "farmer" are made.) Plant genetic resources are spoken of as the basis of food and agricultural systems, farmers' rights are affirmed, and conservation is seen as critical with "local and indigenous communities and farmers, particularly those in the centres of origin and crop diversity" providing an "enormous contribution" to the "conservation and development of plant genetic resources" (FAO 1983: 12). The Plant Breeders' Bill does not mention this specific treaty (five mentions of "treaty"), but "treaty" is spoken about in the context that "a foreign citizen or a resident in the territory of a party to a treaty to which the Republic is party" can apply for a plant breeder right in Ghana (Plant Breeders' Bill 2013: 7).

13 The Peasant Farmers' Association of Ghana notes that the same problematic elements that were a part of the PBB are retained in the Plant Variety Protection Bill, and that parliament did not respond to their concerns (Ghanaweb.com 2020),

14 "Farmer" or "farmers" was used in the search of the document.

15 Interview with official from the Forum for Agricultural Research in Africa (FARA), May 15, 2015.

16 Furthermore, the provision that "a legal entity with its registered office in the country or within the territory of a State which is a party to an international treaty to which the Republic is a party may also file an application" (Plant Breeders' Bill 2013: 3) means that US legal entities could file for plant breeder rights if both Ghana and the United States ratify the International Treaty on Plant Genetic Resources for Food and Agriculture.

17 The Plant Variety Protection Bill passed by parliament retains a similar provision.

18 At the time the writ of summons was filed in February 2015, there was not yet a National Biosafety Authority but rather a National Biosafety Committee years after the passage of the Biosafety Act in December 2011. Food Sovereignty Ghana likely thought that this could be a way to advance the case since the government was apparently not abiding by its own legislation. The National Biosafety Authority was inaugurated on the same date that the court case began.

19 As discussed in the previous chapter, the production of foundation seed was previously the exclusive domain of the public sector, subject to national registration with a regulatory board (Kuhlmann and Zhou 2016: 6; AGRA-SSTP n.d.).

20 Although the point behind the slogan "Food sovereignty cools the planet"— recognizing how agroecological practices embraced by the food sovereignty movement have great climate mitigating potential—is, of course, well taken.

21 Interview, June 28, 2017.

22 Interview, July 19, 2017.

23 An important exception was the FAO representative present.

24 Interview with a senior advisor to PBS, May 15, 2015.

25 Interview with agricultural research scientist at SARI, February 13, 2013.

26 Interview with agricultural research scientist at SARI, May 7, 2015.

27 Interview with members of the National MoFA board of directors, March 25, 2013.

28 Interview with agricultural research scientist at SARI, May 7, 2015. As I have shown in the book, the ability of farmers to make such a choice is not a neutral technocratic issue but is rather contingent on access to techno-scientific knowledge and expertise that most farmers, and even most elected officials, in Ghana lack. In this statement, farmers are conceived of as consumers whose democratic participation is reduced to a vote in the marketplace.

29 Conversations with members of Food Sovereignty Ghana, May 5, 2015; May 12, 2015; and August 7, 2018.

30 Interview with agricultural research scientist at SARI, February 13, 2013.

31 Interview with acting CEO of National Biosafety Authority, May 20, 2015.

32 It should be stressed that the debate over GMOs in Ghana is in an early stage, and this is a period of transition. However, observations at this stage indicate that many of the dynamics I have seen in South Africa and around the world, theorized in Chapter 2, are likely to also play out in Ghana.

33 Several of my Ghanaian informants suggested that there were questions about the authenticity of this organization, suggesting it was not a grassroots farmers' and fisherfolk association. Such suggestions were lent credibility when I interviewed a GNAFF spokesman on May 15, 2015, about whether the organization had any concerns about GMOs. He did not answer the question but rather referred back to the opportunity he had in September 2014 to travel to the Monsanto-funded Danforth Plant Center and emphasized the importance of practicing "modern farming."

34 Interview with agricultural biotechnology expert, March 6, 2013.

35 As discussed in Chapter 3, intellectual property rights governing plants (as embodied in the WTO TRIPS Agreement) have often obscured the prior contributions of indigenous and traditional uses of plants, rather treating them differentially as the "common heritage of mankind" and not intellectual property (e.g., Mgbeoji 2006; Shiva 2000; Bratspies 2007).

36 Interview, August 6, 2018.

37 Dr. Rita Mumm and Dr. Vernon Gracen are examples of US academics with agribusiness industry ties who have played important roles in shaping the WACCI curriculum.

Conclusion

1 Such as the Ghana Plants and Fertilizer Act 2010 and the UPOV-compliant Plant Variety Protection Bill that passed in parliament on October 9, 2020.

2 That connect to "last mile users" and integrate farmers into private farmers' associations, as discussed in Chapter 5.

3 From films screened by projector backpacks in villages across northern Ghana to graduate curriculum at the West African Centre for Crop Improvement.

4 As stated by Kwesi Ahwoi, former minister of food and agriculture in Ghana, June 27, 2011.

5 Take, for example, a grant from the BMGF to a Denver-based company to develop a *gari* roaster—used to process cassava into the delicious condiment *gari*—rather than fund the many "makers" that shine at the impressive annual Maker's Faire in Nigeria to produce one.

6 AGRA's work to support biotechnology is more subtle and behind the scenes. As one AGRA official told me with respect to the Plant Breeders' Bill that would create the "enabling environment" for foreign plant breeders, "AGRA is not taking the lead, MoFA is, because we are an external entity," adding, "it is getting too political" (conversation with AGRA official, June 29, 2017). Review of BMGF grant database with search "agriculture Ghana" (206 results), no date restrictions, https://www.gatesfoundation. org/How-We-Work/Quick-Links/Grants-Database, accessed January 22, 2020.

7 Interview, May 4, 2012.

8 Ibid.

9 And a further reason why Monsanto agreed to remove its name following Bayer's acquisition of the company.

10 I have been on the Monsanto listserv since 2009 and note that insect resistance is of constant concern in their product development strategies.

11 The Nagoya Protocol is a supplemental agreement to the Convention on Biological Diversity with 123 parties and 50 countries that have ratified. Ghana has ratified; the United States has not signed (https://www.cbd.int/abs/nagoya-protocol/signatories/, accessed December 21, 2020).

12 Richards (1985: 155) writes, "Under the concept of 'people's science' the 'agricultural expert' is replaced by the notion of an agent who is a catalyst and facilitator."

BIBLIOGRAPHY

3ADI (African Agribusiness and Agro-Industries Development Initiative) (n.d.) "Why 3ADI?," http://www.3adi.org/, accessed July 7, 2015.

AATF (African Agricultural Technology Foundation) (n.d.a) "*Maruca*-resistant cowpea: Frequently asked questions," http://www.aatf-africa.org/userfiles/CowpeaFAQ. pdf, accessed July 5, 2015.

AATF (n.d.b) "Open Forum on Agricultural Biotechnology in Africa (OFAB)," http://www.aatf-africa.org/projects-programmes/programmes/open-forum-agricultural-biotechnology-africa-ofab, accessed January 23, 2016.

AATF (n.d.c) "Our donors," http://aatf-africa.org/about-us/governance/our-donors, accessed July 3, 2015.

AATF (n.d.d) "Nitrogen-use efficient, water-use efficient, and salt-tolerant rice project," http://www.aatf-africa.org/files/Rice-project-brief.pdf, accessed July 5, 2015.

AATF (n.d.e) "Researchers inventing pod borer resistant cowpea for Africa," http://cowpea.aatf-africa.org/news/media/researchers-inventing-pod-borer-resistant-cowpea-africa, accessed July 5, 2015.

Abdulai, A., and W. Huffman (2000) "Structural adjustment and economic efficiency of rice farmers in Northern Ghana," *Economic Development and Cultural Change*, 48(3): 503–20.

ACDI/VOCA (Agricultural Cooperative Development International and Volunteers in Overseas Cooperative Assistance) (n.d.) "Farming as a business: Development tool promotes both food and income security," http://www.acdivoca.org/farming-as-a-business, accessed July 7, 2015.

Adati, T., M. Tamò, S. R. Yusuf, and W. Hammond (2007) "Integrated pest management for cowpea–cereal cropping systems in the West African savannah," *International Journal of Tropical Insect Science*, 27(3–4): 123–37.

Adino, S., Z. Wondifraw, and M. Addis (2018) "Replacement of soybean grain with cowpea grain (*Vigna unguiculata*) as protein supplement in Sasso x Rir crossbred chicks diet," *Poultry, Fisheries & Wildlife Sciences*, 6(1): 1–6.

Africa Lead Team II (2018) *Feed the Future: Building Capacity for African Agricultural Transformation (Africa Lead II)*, Washington, DC: USAID Bureau of Food Security.

African Centre for Biosafety (2009) *Africa Bullied to Grow Defective Bt Maize: The Failure of Monsanto's MON810 Maize in South Africa*, Melville, South Africa: African Centre for Biosafety.

AFSA (Alliance for Food Sovereignty in Africa) (2014) *Comments on the FAO Draft Guide for National Seed Policy Implementation*, Kampala: Alliance for Food Sovereignty in Africa.

AFSA (n.d.) "What is AFSA," http://afsafrica.org/what-is-afsa/, accessed July 3, 2015.

Agarwal, B. (2014) "Food sovereignty, food security and democratic choice: Critical contradictions, difficult conciliations," *Journal of Peasant Studies*, 41(6): 1247–68.

AGRA (Alliance for a Green Revolution in Africa) (2013) *AGRA in 2012: Moving from Strength to Strength*, Nairobi, Kenya: AGRA.

AGRA (2014) *Africa Agriculture Status Report 2014: Climate Change and Smallholder Agriculture in Sub-Saharan Africa*, Nairobi, Kenya: AGRA.

AGRA (2019) "Seed legislation: Nigeria takes steps to secure the legacy of plant breeders and foundation seed producers," March 18, https://agra.org/seed-legislation-nigeria-takes-steps-to-secure-the-legacy-of-plant-breeders-and-foundation-seed-producers/, accessed December 18, 2020.

AGRA (2015) *Progress Report 2007–2014*, Nairobi, Kenya: AGRA.

AGRA (n.d.a) "The African Seed Company toolbox," http://www.agra.org/agra/en/what-we-do/the-african-seed-company-toolbox/, accessed July 6, 2015.

AGRA (n.d.b) "AGRA grantee wins national award in seed production in Ghana: Award expected to strengthen public-private partnerships," http://archive.agra.org/media-centre/news/agra-grantee-wins-national-award-in-seed-production-in-ghana/, accessed July 7, 2015.

AGRA (n.d.c) "AGRA: Growing Africa's agriculture," http://agra-alliance.org/, accessed July 3, 2015.

AGRA (n.d.d) "Where we work," http://agra-alliance.org/where-we-work/where-we-work/#top-link, accessed July 7, 2015.

AGRA (n.d.e) "Who we are: History of AGRA," http://www.agra.org/who-we-are/our-history/, accessed July 3, 2015.

AGRA (n.d.f) "Ghanaian president hosts AGRA board members at state house, Ghana," http://agra-alliance.org/media-centre/news/president-mahama-meets-agra-board/, accessed July 7, 2015.

AGRA-SSTP (Alliance for a Green Revolution in Africa-Scaling Seeds and Technologies Partnership) (n.d.) *Ghana Early Generation Seed Study*, Washington, DC: USAID.

AGRA Watch (2010) "For immediate release: Gates Foundation invests in Monsanto," *AGRA Watch Press Release*, August 25, https://cagj.org/2010/08/for-immediate-release-gates-foundation-invests-in-monsanto/, accessed January 22, 2020.

AGRA Watch (n.d.) "Monitoring the Gates Foundation and AGRA, promoting food sovereignty and agricultural sustainability in Africa," http://www.cagj.org//wp-content/uploads/AWbrochure.pdf, accessed July 7, 2015.

AgroNews (2012) "Kenya banned importation of all GMO Foods," *AgroNews*, http://news.agropages.com/News/NewsDetail---8425.htm, accessed July 3, 2015.

Ahwoi, K. (2011) "Statement by honourable Kwesi Ahwoi at FAO," government of Ghana, statement at the 37th session of the FAO Conference, Rome, June 27, http://www.ghana.gov.gh, accessed November 11, 2011.

Aistara, G. A. (2012) "Privately public seeds: Competing visions of property, personhood, and democracy in Costa Rica's entry into CAFTA and the Union for Plant Variety Protection," *Journal of Political Ecology*, 19: 127–44.

Akintola, A., B. Matthias, and C. Hardcastle. (2003) *Public-Private Partnerships: Managing Risks and Opportunities*, Oxford: Blackwell Science.

Akologo-Azupogo, H. (2018) "Commoditization of Land and Its Implications for the Livelihood Security of the Upper East Region of Ghana," doctoral dissertation, Millar Institute for Transdisciplinary and Development Studies, Bolgatanga, Ghana.

Akram-Lodhi, A. H. (2013) *Hungry for Change: Farmers, Food Justice and the Agrarian Question*, Halifax: Fernwood.

Akram-Lodhi, A. H., and C. Kay (2010) "Surveying the agrarian question (part 2): Current debates and beyond," *Journal of Peasant Studies*, 37(2): 255–84.

Alhassan, W. (2001) *The Status of Agricultural Biotechnology in Selected West and Central African Countries*, Ibadan, Nigeria: International Institute of Tropical Agriculture.

Alhassan, W. (2010) "Agricultural biotechnologies in developing countries: Options and opportunities in crops, forestry, livestock, fisheries and agro-industry to face the challenges of food insecurity and climate change," Forum for Agricultural Research in Africa (FARA) paper presented at the UN Food and Agriculture Organization Technical Conference on Agricultural Biotechnologies in Developing Countries, Guadalajara, Mexico, March 1–4.

Allain, R. (2005) "How do you define a vector?" Wired, June 24, https://www.wired.com/2015/06/define-vector/, accessed December 21, 2020.

Altieri, M. A. (1999) "The ecological role of biodiversity in agroecosystems," *Agriculture, Ecosystems and Environment*, 74: 19–31.

Altieri, M. A. (2004) *Genetic Engineering in Agriculture: The Myths, Environmental Risks, and Alternatives*, 2nd ed., Oakland: Food First Books.

Altieri, M. A., and V. M. Toledo. (2011) "The agroecological revolution in Latin America: Rescuing nature, ensuring food sovereignty and empowering peasants," *Journal of Peasant Studies*, 38(3): 587–612.

Amanor, K. S. (2008) "The changing face of customary land tenure," in *Contesting Land and Custom in Ghana*, J. M. Ubink and K. Amanor (eds.), pp. 55–88, Leiden: Leiden University Press.

Amanor, K. S. (2009) "Global food chains, African smallholders and World Bank governance," *Journal of Agrarian Change*, 9(2): 247–62.

Amanor, K. S. (2011) "From farmer participation to pro-poor seed markets: The political economy of commercial seed networks in Ghana," *IDS Bulletin*, 42(4): 48–58.

Amanor, K. S. (2013) "Chinese and Brazilian cooperation with African agriculture: The case of Ghana," *Future Agricultures Working Paper* 052, http://www.future-agricultures.org/publications/search-publications/political-economy-conference-2013/conference-papers-political-economy-2013/brazilian-and-chinese-engagement/1667-chinese-and-brazilian-cooperation-with-african-agriculture-the-case-of-ghana-1?highlight=WyJhbbWFub3IiXQ==, accessed July 3, 2015.

Amanor, K. S. (2018) "Globalization, agribusiness and the liberalization of agricultural services in Ghana," in *Globalization and Agriculture: Redefining Unequal Development*, A. M. Buanain, M. Rocha de Sousa, and Z. Navarro (eds.), pp. 207–28, London: Lexington Books.

American Soybean Association (2018) "ASA/WISHH grows demand for soy-based poultry feed in Ghana," February 15, https://soygrowers.com/news-releases/asawishh-grows-demand-soy-based-poultry-feed-ghana/, accessed August 7, 2020.

Andrée, P. (2005) "The Cartagena Protocol on Biosafety and shifts in the discourse of precaution," *Global Environmental Politics*, 5(4): 25–46.

Andrée, P. (2007) *Genetically Modified Diplomacy: The Global Politics of Agricultural Biotechnology and the Environment*, Vancouver: UBC Press.

Andrée, P., A. Jeffrey, and M. Bosia (eds.) (2014) *Globalization and Food Sovereignty: Global and Local Change in the New Politics of Food*, Toronto: University of Toronto.

Arrighi, G. (1999) "Globalization, state sovereignty, and the 'endless' accumulation of capital," in *States and Sovereignty in the Global Economy*, D. A. Smith, D. J. Solinger, and S. C. Topik (eds.), pp. 53–73, London: Routledge.

Avant, D. A., M. Finnemore, and S. K. Sell (eds.) (2011) *Who Governs the Globe?*, Cambridge: Cambridge University Press.

Awuni, M. A. (2014) "SADA broke procurement laws in award of 32.4 million afforestation contract to ACI-audit report," *MyJoyOnline*, March 16, http://www.myjoyonline.com/news/2014/april-16th/sada-broke-procurement-laws-in-award-of-324-million-afforestation-contract-to-aci-audit-report.php#sthash.rhBbDDCE. FsJtYGbw.dpuf, accessed July 7, 2015.

Ayele, S. (2007) "The legitimation of GMO governance in Africa," *Science and Public Policy*, 34(4): 239–49.

Ayenan, M., and A. Danquah (2015) "Implications of Plant Breeder Bill on breeding activities in Ghana: Prospective analysis," 5è Colloque des Sciences, Cultures et Technologies de l'Universite d'Abomey-Calavi, September 2015, Abomey-Calavi, Benin.

Banerjee, A. V., and E. Duflo (2011) *Poor Economics: A Radical Rethinking of the Way to Fight Global Poverty*, New York: PublicAffairs.

Barnett, M. (2011) *Empire of Humanity: A History of Humanitarianism*, Ithaca: Cornell University Press.

Basu, P., and B. A. Sholten (2012) "Technological and social dimensions of the Green Revolution: Connecting pasts and futures," *International Journal of Agricultural Sustainability*, 10(2): 109–16.

Bates, R. H. (2005) *Markets and States in Tropical Africa: The Political Basis of Agricultural Policies*, 2nd ed., Berkeley: University of California Press.

Bayard de Volo, L., and E. Schatz. (2004) "From the inside out: Ethnographic methods in political research," *Political Science and Politics*, 37(2): 267–71.

Bell, S. E., A. Hullinger, and L. Brislen (2015) "Manipulated masculinities: Agribusiness, deskilling and the rise of the businessman-farmer in the United States," *Rural Sociology*, 80(3): 285–313.

Bello, W. (2009) *The Food Wars*, New York: Verso.

Benin, S. (2015) "Impact of Ghana's agricultural mechanization services center program," *Agricultural Economics*, 46(S1): 103–17.

Benneh, G. (2011) *Technology Should Seek Tradition: Studies on Traditional Land Tenure and Smallholder Farming Systems in Ghana*, Accra: Ghana Universities Press.

Bennett, J. (2010) *Vibrant Matter: A Political Ecology of Things*, Durham: Duke University Press.

Bernard, B., and A. Lux (2017) "How to feed the world sustainably: An overview of the discourse on agroecology and sustainable intensification," *Regional Environmental Change*, 17(5): 1279–90.

Bernstein, H. (2010) *Class Dynamics of Agrarian Change*, Halifax: Fernwood.

Berry, S. (1993) *No Condition Is Permanent: The Social Dynamics of Agrarian Change in Sub-Saharan Africa*, Madison: University of Wisconsin Press.

Berry, S. (2002) "Debating the land question in Africa," *Comparative Studies in Society and History*, 44(4): 638–68.

Berry, S. (2009) "Property, authority and citizenship: Land claims, politics and the dynamics of social division in West Africa," *Development and Change*, 40(1): 23–45.

Berry, W. (1996) *The Unsettling of America: Culture and Agriculture*, 3rd ed., San Francisco: Sierra Club Books.

Bill & Melinda Gates Foundation (2006) "Bill & Melinda Gates, Rockefeller Foundations form alliance to spur 'Green Revolution' in Africa," September, https://www. gatesfoundation.org/Media-Center/Press-Releases/2006/09/Foundations-Form-Alliance-to-Help-Spur-Green-Revolution-in-Africa, accessed August 3, 2020.

Bill & Melinda Gates Foundation and United States Agency for International Development (2015) "Early generation seed study," https://docs.gatesfoundation.

org/documents/BMGF%20and%20USAID%20EGS%20Study%20Full%20Deck.pdf, accessed August 27, 2019.

Bingen, J. (2008) "Genetically engineered cotton: Politics, science, and power in West Africa," in *Hanging by a Thread: Cotton, Globalization, and Poverty in Africa*, W. G. Moseley and L. C. Gray (eds.), pp. 227–50, Uppsala: Ohio University Press.

Biosorghum (n.d.a) "The ABS Project Consortium," http://biosorghum.org/abs_consort.php, accessed July 5, 2015.

Biosorghum (n.d.b) "The ABS Project Intellectual Property Management Group," http://biosorghum.org/intelectual_property.php, accessed July 5, 2015.

Biosorghum (n.d.c) "ABS Project Development," http://biosorghum.org/abs_consort.php, accessed July 5, 2015.

Birch, K. (2017) "Rethinking value in the bio-economy: Finance, assetization, and the management of value," *Science, Technology, & Human Values*, 42(3): 460–90.

Birch, K., and D. Tyfield (2013) "Theorizing the bioeconomy: Biovalue, biocapital, bioeconomics or … what?" *Science, Technology, & Human Values*, 38(3): 299–327.

Bishop, M., and M. Green. (2009) *Philanthrocapitalism: How Giving Can Save the World*, originally published in 2008 with a new foreword by Bill Clinton in 2009, New York: Bloomsbury.

Blaikie, P. (1985) *The Political Economy of Soil Erosion in Developing Countries*, London: Routledge.

Bob, C. (2005) *The Marketing of Rebellion: Insurgents, Media, and International Activism*, Cambridge: Cambridge University Press.

Bonneuil, C., P. Joly, and C. Marris (2008) "Disentrenching experiment: The construction of GM-crop field trials as a social problem," *Science, Technology, and Values*, 33(2): 201–29.

Borlaug, N., and J. Carter (2008) "Foreword," in *Starved for Science: How Biotechnology Is Being Kept Out of Africa*, R. Paarlberg (ed.), pp. vii–x, Cambridge: Harvard University Press.

Borras, S. M., R. Hall, I. Scoones, B. White, and W. Wolford (2011) "Towards a better understanding of global land grabbing: An editorial introduction," *Journal of Peasant Studies*, 38(2): 209–16.

Boseley, S. (2007) "Gates Foundation may shift billions into ethical stocks after attack on Investments," *Guardian*, January 12.

Bosworth, D. (2011) "The cultural contradictions of philanthrocapitalism," *Society*, 48(5): 382–8.

Boyle, J. (2003) "The second enclosure movement and the construction of the public domain," *Law and Contemporary Problems*, 66(1/2): 33–74.

Bratspies, R. M. (2007) "The new discovery doctrine: Some thoughts on property rights and traditional knowledge," *American Indian Law Review*, 31(2): 315–40.

Bratton, M. (1989) "The politics of government-NGO relations in Africa," *World Development*, 17(4): 569–87.

Bray, C. (2015) "Syngenta chairman sets criteria for further Monsanto talks," *New York Times*, June 23.

Breen, S. (2015) "Saving seeds: The Svalbard Global Seed Vault, Native American seed savers, and problems of property," *Journal of Agriculture, Food Systems, and Community Development*, 5(2): 39–52.

Briggs, H. (2017) "Fall armyworm 'threatens African farmers' livelihoods,'" *BBC News*, February 6.

Brody, J. A. (2015) "Fears, not facts, support G.M.O. labeling," *New York Times*, June 8.

Bromley, S. (1995) "Making sense of structural adjustment," *Review of African Political Economy*, 22(65): 339–48.

Brooks, S. (2005) "Biotechnology and the politics of truth: From the Green Revolution to an evergreen revolution," *Sociologia Ruralis*, 45(4): 360–79.

Brown, N. (2003) "Hope against hype: Accountability in biopasts, presents and futures," *Science Studies*, 16(2): 3–21.

Brown, N., B. Rappert, and A. Webster (eds.) (2000a) *Contested Futures: A Sociology of Prospective Techno-Science*, Aldershot: Ashgate.

Brown, N., B. Rappert, and A. Webster (eds.) (2000b) "Introducing *Contested Futures*: From *looking into* the future to *looking at* the future," in *Contested Futures: A Sociology of Prospective Techno-Science*, N. Brown, B. Rappert, and A. Webster (eds.), pp. 3–20, Aldershot: Ashgate.

Burke, J. (2010) "India to rule on future of Aubergine as country's first genetically modified food," *Guardian*, February 8.

Calhoun, C. (2004) "A world of emergencies: Fear, intervention, and the limits of a cosmopolitan order," *Canadian Review of Sociology and Anthropology*, 41(4): 373–95.

Carney, J. A., and R. N. Rosomoff (2009) *In the Shadow of Slavery: Africa's Botanical Legacy in the Atlantic World*, Berkeley: University of California Press.

CBD (Convention on Biological Diversity) (n.d.) "The Cartagena Protocol on Biosafety," https://bch.cbd.int/protocol, accessed July 3, 2015.

Center for Food Safety & Save Our Seeds (2013) *Seed Giants vs. U.S. Farmers*, Washington, DC: Center for Food Safety.

CGIAR Secretariat (1992) "Agenda item 8—intellectual property rights issues," May 18–22, Mid-Term Meeting, Istanbul, Turkey.

CGIAR Secretariat (2012) "CGIAR principles on the management of intellectual assets," March 7, https://library.cgiar.org/bitstream/handle/10947/3755/CGIAR%20IA%20Principles.pdf?sequence=1, accessed July 8, 2015.

Chalfin, B. (2004) *Shea Butter Republic: State Power, Global Markets, and the Making of an Indigenous Commodity*, New York: Routledge.

Chambers, J. A., P. Zambrano, J. Falck-Zepeda, G. Gruère, D. Sengupta, and K. Hokanson (2014) *GM Agricultural Technologies for Africa*, Washington, DC: International Food Policy Research Institute.

Chambers, R. (1983) *Rural Development: Putting the Last First*, London: Routledge.

Chambers, R., and G. R. Conway (1991) "Sustainable rural livelihoods: Practical concepts for the 21st century," *IDS Discussion Paper*, 296, Brighton: IDS.

Chouliaraki, L. (2013) *The Ironic Spectator: Solidarity in the Age of Post-Humanitarianism*, Cambridge, MA: Polity.

CIKOD (n.d.) "About us," www.cikodghana.org, accessed July 28, 2020.

Circle of Blue (n.d.) "Global map of 'land grabs' by country and by sector," http://www.circleofblue.org/LAND.html, accessed July 7, 2015.

Clapp, J. (2012a) *Food*, Cambridge: Polity.

Clapp, J. (2012b) *Hunger in the Balance: The New Politics of International Food Aid*, Ithaca: Cornell University Press.

Clapp, J. (2014) "Responsibility to the rescue? Governing private investment in global agriculture," paper presented at the International Studies Association, Toronto, Canada, March 26.

Clapp, J., and D. Fuchs (eds.) (2009) *Corporate Power in Global Agrifood Governance*, Cambridge, MA: MIT Press.

Cole, T. (2012) "The white-savior industrial complex," *Atlantic*, March 21.

Collier, P., and S. Dercon (2014) "African agriculture in 50 years: Smallholders in a rapidly changing world?," *World Development*, 63: 92–101.

Comaroff, J., and J. L. Comaroff (2012) *Theory from the South: Or, How Euro-America Is Evolving Toward Africa*, London: Paradigm.

COMPAS (2006) *Endogenous Development in Practice: Towards Well-Being of People and Ecosystems*, Leusden, Netherlands: ETC Compas.

Conway, G. (1997) *A Doubly Green Revolution: Food for All in the Twenty-First Century*, London: Penguin Books.

Cooper, F. (2002) *Africa since 1940: The Past of the Present*, Cambridge: Cambridge University Press.

Cooper, M. (2008) *Life as Surplus: Biotechnology and Capitalism in the Neoliberal Era*, Seattle: University of Washington Press.

Cooke, J. G., and K. Flowers (2018) *Feed the Future in Ghana: Promising Progress, Choices Ahead*, Washington, DC: Center for Strategic and International Studies.

Cotula, L. (2013) *The Great African Land Grab? Agricultural Investments and the Global Food System*, London: Zed Books.

Curtis, M., and D. Adama (2013) *Walking the Walk: Why and How African Governments Should Transform Their Agriculture Spending*, Washington, DC: ActionAid.

Cutler, A. C., T. Porter, and V. Haufler (eds.) (1999) *Private Authority and International Affairs*, Albany: SUNY.

Daño, E. C. (2007) *Unmasking the New Green Revolution in Africa: Motives, Players and Dynamics*, Penang, Malaysia; Bonn, Germany; and Richmond, South Africa: Joint publication by Third World Network, Church Development Service, and the African Centre for Biosafety.

Declaration of Nyéléni (2007) Sélingué, Mali, February 27, https://nyeleni.org/spip. php?article290, accessed December 16, 2020.

Dempsey, J. (2016) *Enterprising Nature: Economics, Markets, and Finance in Global Biodiversity Politics*, West Sussex, UK: Wiley.

Desmond, E. (2016) "The legitimation of development and GM crops: The case of Bt cotton and indebtedness in Telangana, India," *World Development Perspectives*, 1(March): 23–5.

Djurfeldt, G., H. Holmén, M. Jirstrøm, and R. Larsson (eds.) (2005) *The African Food Crisis: Lessons from the Asian Green Revolution*, Cambridge: CAB International.

Donald Danforth Plant Science Center (n.d.) "BioCassava Plus," https://www. danforthcenter.org/scientists-research/research-institutes/institute-for-international-crop-improvement/crop-improvement-projects/biocassava-plus, accessed July 5, 2015.

Dowd-Uribe, B. M. (2014) "Engineering yield and inequality? How institutions and agro-ecology shape Bt cotton outcomes in Burkina Faso," *Geoforum*, 53: 161–71.

Dowd-Uribe, B. M., and J. Bingen (2011) "Debating the merits of biotech crop adoption in sub-Saharan Africa: Distributional impacts, climatic variability and pest dynamics," *Progress in Development Studies*, 11(1): 63–8.

Downs, A. (1972) "Up and down with ecology: The 'issue-attention cycle,'" *Public Interest*, 28: 38–50.

Dubb, A. (2018) "The value components of contract farming in contemporary capitalism," *Journal of Agrarian Change*, 18(4): 882–92.

Du Bois, C. M., and I. Sergio Freire de Sousa (2008) "Genetically engineered soy," in *The World of Soy*, C. M. DuBois, C. Tan, and S. Mintz (eds.), pp. 74–96, Urbana: University of Illinois Press.

Dupoix, P., T. Ermias, S. Huezé, S. Niavas, and M. Von Koschitzky Kimani (2014) *Winning in Africa: From Trading Posts to Ecosystems*, Boston: Boston Consulting Group on behalf of Private Equity Africa, http://www.privateequityafrica.com/wpm/wp-content/uploads/2014/03/2014_January-_Winning_in_Africa_BCG.pdf, accessed July 6, 2015.

DuPont Pioneer (2013) "Media statement: DuPont Pioneer completes acquisition of Pannar Seed," October 29, http://www.pioneer.com/home/site/about/news-media/news-releases/template.CONTENT/guid.2A29ED4E-7EE4-A71B-46BB-49973731A7A1, accessed July 6, 2015.

Easterly, W. (2007) *The White Man's Burden: Why the West's Efforts to Aid the Rest Have Done So Much Ill and So Little Good*, New York: Penguin.

Economist (2006) "The birth of philanthrocapitalism," February 25.

Edelman, M. (2014) "Food sovereignty: Forgotten genealogies and future regulatory challenges," *Journal of Peasant Studies*, 41(6): 959–78.

Edelman, M., T. Weis, A. Baviskar, S. M. Borras Jr., E. Holt-Giménez, D. Kandiyoti, and W. Wolford (2014) "Introduction: Critical perspectives on food sovereignty," *Journal of Peasant Studies*, 41(6): 911–31.

Edkins, J. (2000) *Whose Hunger? Concepts of Famine, Practices of Aid*, Minneapolis: University of Minnesota Press.

Elgert, L. (2016) "'More soy on fewer farms' in Paraguay: Challenging neoliberal agriculture's claims to sustainability," *Journal of Peasant Studies*, 43(2): 537–61.

Ellis, F. (2000) *Rural Livelihoods and Diversity in Developing Countries*, Oxford: Oxford University Press.

el-Tahiri, J. (2004) *The Price of Aid*, 55 minutes.

Epstein, C. (2005) "Knowledge and power in global environmental activism," *International Journal of Peace Studies*, 10(1): 47–67.

Epstein, C. (2008) *The Power of Words in International Relations: Birth of an Anti-Whaling Discourse*, Cambridge, MA: MIT Press.

Escobar, A. (1995) *Encountering Development: The Making and Unmaking of the Third World*, Princeton, NJ: Princeton University Press.

Escobar, A. (2008) *Territories of Difference: Place, Movements, Life, Redes*, Durham, NC: Duke University Press.

Escobar, A. (2010) "Planning," in *The Development Dictionary: A Guide to Knowledge as Power*, W. Sachs (ed.), 2nd ed., pp. 145–61, London: Zed Books.

ETC Group (n.d.) "Myth: Monsanto sells terminator seeds," http://www.monsanto.com/newsviews/pages/terminator-seeds.aspx, accessed July 6, 2015.

Evenson, R. E., and D. Gollin (2003) "Assessing the impact of the Green Revolution, 1960 to 2000," *Science*, 300(5620): 58–762.

Fairhead, J., M. Leach, and I. Scoones (eds.) (2012) "Green grabbing: A new appropriation of nature?" *Journal of Peasant Studies*, 39(2): 237–61.

Fairhead, J., M. Leach, and I. Scoones (eds.) (2013) *Green Grabbing: A New Appropriation of Nature*, London: Routledge.

Fan, S., B. Omilola, and M. Lambert (2009) "Public spending for agriculture in Africa: Trends and composition," *Regional Strategic Analysis and Knowledge Support System Working Paper No. 28*, Washington, DC: Regional Strategic Analysis and Knowledge Support System.

Fan, S., J. Brzeska, M. Keyzer, and A. Halsema (2013) *From Subsistence to Profit: Transforming Smallholder Farms*, Washington, DC: International Food Policy Research Institute.

FAO (Food and Agriculture Organization of the United Nations) (1983) *Resolution 8/83: International Undertaking on Plant Genetic Resources*, http://www.fao.org/3/AK668E/AK668E.pdf, accessed December 20, 2020.

FAO (2006) "Food security," *FAO Policy Brief*, 2, http://www.fao.org, accessed July 13, 2020.

FAO (2017) "African nations adopt a common position on the International Plant Treaty," March 10, http://www.fao.org/plant-treaty/news/news-detail/en/c/1041549/, accessed July 22, 2020.

FAO (n.d.a) "Comprehensive Africa Agriculture Development Programme (CAADP)," http://www.un.org/en/africa/osaa/peace/caadp.shtml, accessed March 16, 2019.

FAO (n.d.b) "What are good agricultural practices?," http://www.fao.org/3/a-a1193e.pdf, accessed July 6, 2015.

FAO and CABI (2019) *Community-Based Fall Armyworm (Spodoptera Frugiperda) Monitoring, Early Warning and Management: Training of Trainers Manual*, 1st ed., Rome: Food and Agriculture Organization of the United Nations.

FAS/USDA (Foreign Agricultural Service/United States Department of Agriculture) (2002) "USAID announces international biotech collaboration," June 12, http://www.fas.usda.gov/icd/summit/2002/statearchive/USAIDbiotech.htm, accessed November 8, 2011.

FAS/USDA (2010) "Ghana: Biotechnology-GE plants and animals," *Global Agricultural Information Network (GAIN) Report*, July 14.

FAS/USDA (2013a) "Agricultural biotechnology in Mozambique," *Global Agricultural Information Network (GAIN) Report*, November 1.

FAS/USDA (2013b) "Ghana agricultural biotechnology annual," *Global Agricultural Information Network (GAIN) Report*, September 3.

Faulkner, R. (ed.) (2006) *The International Politics of Genetically Modified Food: Diplomacy, Trade and Law*, New York: Palgrave Macmillan.

Feed the Future (2011) "Ghana: FY 2011–2015 multi-year strategy," February 22, US government document, http://feedthefuture.gov/country/ghana, accessed July 7, 2015.

Feed the Future (2018) *Ghana Food Security Strategy (GFSS) Country Plan for Ghana*, August 10, Washington, DC: US Government.

Feed the Future (n.d.) "Feed the Future Ghana: Agriculture Technology Transfer Project—overview and accomplishments," *Briefing Presentation*, USAID and IFDC.

Ferguson, J. (1994) *The Anti-Politics Machine: "Development," Depoliticization, and Bureaucratic Power in Lesotho*, Minneapolis: University of Minnesota Press.

Ferguson, J. (2006) *Global Shadows: Africa in the Neoliberal World Order*, Durham: Duke University Press.

Finger, J. M., and P. Schuler (eds.) (2004) *Poor People's Knowledge: Promoting Intellectual Property in Developing Countries*, Washington, DC: Co-publication of the World Bank and Oxford University Press.

Fitting, E. (2011) *The Struggle for Maize: Campesinos, Workers, and Transgenic Corn in the Mexican Countryside*, Durham: Duke University Press.

Flaherty, K., G. O. Essegbey, and R. Asare (2010) *Ghana: Recent Developments in Agricultural Research*, Rome: Agricultural Science and Technology Indicators.

Flyvbjerg, B. (2001) *Making Social Science Matter: Why Social Inquiry Fails and How It Can Succeed Again*, Cambridge: Cambridge University Press.

Food Sovereignty Ghana (2014a) "COFAM press statement," January 28, https://foodsovereigntyghana.org/category/our-campaigns/national-campaign-against-upov-plant-breeders-bill/, accessed February 2, 2014.

Food Sovereignty Ghana (2014b) "Communiqué: FSG workshop on GMOs, seed laws, and biosafety," February 28, https://foodsovereigntyghana.org/page/6/, accessed January 23, 2016.

Food Sovereignty Ghana (2015) "Food Sovereignty Ghana marks second anniversary," March 24, https://foodsovereigntyghana.org/, accessed May 20, 2015.

Food Sovereignty Ghana (2018) " 'Ghana's Plant Breeders Bill lacks legitimacy! It must be revised!'—CSOs tell Parliament," March 2, https://foodsovereigntyghana.org/, accessed July 22, 2020.

Foresight (2011) *The Future of Food and Farming: Executive Summary*, London: Government Office for Science.

Foucault, M. (1972) *The Archaeology of Knowledge and the Discourse on Language*, A. M. Sheridan Smith (trans.), New York: Pantheon Books.

Foucault, M. (1979) *Discipline and Punish: The Birth of the Prison*, New York: Vintage.

Foucault, M. (1990) *The History of Sexuality, Volume 1: An Introduction*, Robert Hurley (trans.), New York: Vintage Books.

Foucault, M. (2003) *Society Must Be Defended: Lectures at the Collège de France, 1975–1976*, David Macey (trans.), New York: Picador.

Foucault, M. (2008) *The Birth of Biopolitics: Lectures at the Collège de France, 1978–1979*, M. Senellart (ed.) and G. Burchell (trans.), New York: Palgrave Macmillan.

Francis, E. (2000) *Making a Living: Changing Livelihoods in Rural Africa*, London: Routledge.

Freebairn, D. K. (1995) "Did the Green Revolution concentrate incomes? A quantitative study of research reports," *World Development*, 23(2): 265–79.

Friedmann, H. (1993) "The political economy of food: A global crisis," *New Left Review*, 197(1): 29–57.

Fuchs, D. (2005) "Commanding heights? The strength and fragility of business power in global politics," *Millennium*, 33: 771–801.

G8 New Alliance for Food Security and Nutrition (n.d.) *G8 Cooperation Framework to Support the "New Alliance for Food Security and Nutrition" in Ghana*, http://www.state. gov/documents/organization/190626.pdf, accessed July 1, 2015.

GAIN (Global Agricultural Information Network) (2012) "Agricultural news for Italy EU and World July 2012," http://gain.fas.usda.gov/Recent%20GAIN%20Publications/ Agricultural%20News%20for%20Italy%20EU%20and%20World%20January%20 2012_Rome_Italy_2-13-2012.pdf, accessed January 17, 2016.

Gates, B., and M. Gates (2006) "Bill & Melinda Gates, Rockefeller Foundations form alliance to help spur 'Green Revolution' in Africa," *Bill & Melinda Gates Foundation Press Release*, September, https://www.gatesfoundation.org/Media-Center/Press-Releases/2006/09/Foundations-Form-Alliance-to-Help-Spur-Green-Revolution-in-Africa, accessed January 15, 2020.

Gates, B., and M. Gates (2014) "3 myths that block progress for the poor," *The Bill and Melinda Gates Foundation Annual Letter*.

Gathura, G. (2004) "GM technology fails local potatoes," *Daily Nation, Kenya*, January 29.

GBDI (n.d.) "Welcome to GBDI," http://www.gbdi.org/, accessed July 6, 2015.

George, S. (1988) *A Fate Worse Than Debt: A Radical New Analysis of the Third World Debt Crisis*, London: Penguin Books.

Ghana Biosafety Act 831 (2011).

Ghana Commercial Agriculture Project (GCAP) (n.d.) "About us," https://gcap.org.gh/ about-us/, accessed August 21, 2020.

Ghana Patents Act (2003).

Ghanaweb.com (2020) "Why peasant farmers are against newly passed Plant Variety Protection Bill," November 6, https://www.ghanaweb.com/GhanaHomePage/

business/Why-peasant-farmers-are-against-newly-passed-Plant-Variety-Protection-Bill-1102486, accessed December 21, 2020.

Gibbon, P. (1992) "A failed agenda? African agriculture under structural adjustment with specific reference to Kenya and Ghana," *Journal of Peasant Studies*, 20(1): 50–96.

Gibbon, P., and S. Ponte (2005) *Trading Down: Africa, Value Chains, and the Global Economy*, Philadelphia: Temple University Press.

GINN (Global Impact Investment Network) (n.d.) "About the Global Impact Investment Network (GIIN)," http://www.thegiin.org/cgi-bin/iowa/resources/about/index.html, accessed July 1, 2015.

Girdner, J., V. Olorunsola, M. Froning, and E. Hansen (1980) "Ghana's agricultural food policy: Operation Feed Yourself," *Food Policy* 5(1): 14–25.

Glover, D. (2010a) "The corporate shaping of GM crops as a technology for the poor," *Journal of Peasant Studies*, 37(1): 67–90.

Glover, D. (2010b) "Is *Bt* cotton a pro-poor technology? A review and critique of the empirical record," *Journal of Agrarian Change*, 10(4): 482–509.

GMO Awareness (n.d.) "Movies to watch—GMO and more," http://gmo-awareness.com/resources/movies-to-watch-gmo-and-more/, accessed July 3, 2015.

Goldman, A., and J. Smith (1995) "Agricultural transformations in India and northern Nigeria: Exploring the nature of Green Revolutions," *World Development*, 23(2): 243–63.

Goldman, M. (2005) *Imperial Nature: The World Bank and Struggles for Social Justice in the Age of Globalization*, New Haven, CT: Yale University Press.

Gómez, J. et al. (2013) "Microencapsulated *Spodoptera frugiperda* nucleopolyhedrovirus: Insecticidal activity and effect on arthropod populations in maize," *Biocontrol Science and Technology*, 23(7): 829–46.

Government of Ghana (2011) "Statement by honourable Kwesi Ahwoi at FAO," at the 37th session of the FAO Conference, Rome, June 27, http://www.ghana.gov.gh/index.php?option=com_content&view=article&id=6412:state ment-of-honourable-kwesi-ahwoi-minister-for-food-and-agriculture-ghana-at-the-thirty-seventh-session-of-the-fao-conference-held-in-rome-italy-25-june--2-july-2011&catid=56:speeches&Itemid=205, accessed November 11, 2011.

Government of Ghana (n.d.) "About Ghana: Regions," http://www.ghana.gov.gh, accessed July 2, 2015.

Graddy-Lovelace, G. (2014) "Situating in situ: A critical geography of agricultural biodiversity conservation in the Peruvian Andes and beyond," *Antipode*, 46(2): 426–54.

Graddy-Lovelace, G. (2020) "Plants: Crop diversity pre-breeding technologies as agrarian care co-opted?" *Royal Geographical Society*, 52(2): 235–43.

GRAIN (2014a) *How Does the Gates Foundation Spend Its Money to Feed the World?*, Barcelona: GRAIN.

GRAIN (2014b) "Hungry for land: Small farmers feed the world—with less than a quarter of all farmland," *GRAIN Press Release*, May 29, https://www.grain.org/article/entries/4929-hungry-for-land-small-farmers-feed-the-world-with-less-than-a-quarter-of-all-farmland, accessed July 7, 2015.

GRAIN (2015) *Profiting from Climate Crisis, Undermining Resilience in Africa: Gates and Monsanto's Water-Efficient Maize for Africa (WEMA) Project*, May 4, Barcelona: GRAIN.

GRAIN (n.d.) *Don't Get Fooled Again! Unmasking Two Decades of Lies About Golden Rice*, Barcelona: GRAIN.

Greenpeace (2015) "Greenpeace media briefing: EU parliament to adopt new GM crop national opt-out law," January 12, http://www.greenpeace.org/eu-unit/Global/eu-unit/reports-briefings/2015/GMOs%20briefing%2012012015%20%20FINAL.pdf, accessed July 3, 2015.

Griffin, K. (1974) *The Political Economy of Agrarian Change: An Essay on the Green Revolution*, London: Macmillan.

Grove, R. H. (1996) *Green Imperialism: Colonial Expansion, Tropical Island Edens and the Origins of Environmentalism, 1600–1860*, Cambridge: Cambridge University Press.

Grow Africa (n.d.) "Ghana: Ghana Commercial Agriculture Project," http://growafrica.com/initiative/ghana, accessed July 7, 2015.

Grow Africa Secretariat (2013) *Grow Africa: Investing in the Future of African Agriculture, 1st Annual Report*, Geneva, Switzerland.

Gunset, G. (1999) "DuPont to buy Pioneer Hi-Bred as agribusiness mergers heat up," *Chicago Tribune*, March 16.

Gustafson, D., E. Jimenez, and J. F. Linn (1989) *Developing the Private Sector: A Challenge for the World Bank Group*, report no. 10378, Washington, DC: World Bank Group.

Gyasi, E. A., and J. I. Uitto (eds.) (1997) *Environment, Biodiversity and Agricultural Change in West Africa*, New York: United Nations University Press.

Hall, R., I. Scoones, and D. Tsikata (2017) "Plantations, outgrowers and commercial farming in Africa: Agricultural commercialisation and implications for agrarian change," *Journal of Peasant Studies*, 44(3): 515–37.

Harmon, A. (2013) "Golden Rice: Lifesaver?," *New York Times*, August 24.

Hartwich, F., J. Tola, A. Engler, C. González, G. Ghezan, J. M. P. Vázquez-Alvarado, J. A. Silva, J. d. J. Espinoza, and M. V. Gottret (2008) *Building Public-Private Partnerships for Agricultural Innovation*, Washington, DC: IFPRI.

Harvard College v Canada (Commissioner of Patents) (2002) 4 S.C.R. 45, 2002 SCC 76, Supreme Court of Canada.

Harvey, D. (2004) "The 'new' imperialism: Accumulation by dispossession," *Socialist Register*, 40: 63–87.

Harvey, D. (2005) *A Brief History of Neoliberalism*, New York: Oxford University Press.

Harvey, D. (2006) *Spaces of Global Capitalism: Towards a Theory of Uneven Geographical Development*, London: Verso.

Hastert, D. J. (2003) *Plant Biotechnology Research and Development in Africa: Challenges and Opportunities, Hearing before the Subcommittee on Research Committee on Science House of Representatives*, 108th Cong. statement, June 12, p. 11.

Haufler, V. (2003) "Globalization and industry self-regulation," in *Governance in a Global Economy: Political Authority in Transition*, M. Kahler and D. Lake (eds.), pp. 226–54, Princeton: Princeton University Press.

Haufler, V. (2009) "The Kimberley Process certification scheme: An innovation in global governance and conflict prevention," *Journal of Business Ethics*, 89(4): 403–16.

Haverkort, B., D. Millar, and C. Gonese (2003) "Knowledge and belief systems in sub-Saharan Africa," in *Ancient Roots, New Shoots: Endogenous Development in Practice*, B. Haverkort, K. van 't Hooft, and W. Hiemstra (eds.), pp. 29–36, London: Zed Books.

Heller, C. (2013) *Food, Farms, and Solidarity*, Durham: Duke University Press.

Helmreich, S. (2008) "Species of biocapital," *Science as Culture*, 17(4): 463–78.

Henrich, J. (2001) "Cultural transmission and the diffusion of innovations: Adoption dynamics indicate that biased cultural transmission is the predominate force in behavioral change," *American Anthropologist*, 103(4): 992–1013.

Herring, R. (2008) "Opposition to transgenic technologies: Ideology, interests and collective action frames," *Nature Reviews Genetics*, 9(6): 458–63.

Herring, R. (2010) "Framing the GMO: Epistemic brokers, authoritative knowledge, and diffusion of opposition to biotechnology," in *The Diffusion of Social Movements: Actors, Mechanisms, and Political Effects*, R. Givan, K. Roberts, and S. Soule (eds.), Cambridge: Cambridge University Press.

Hill, C. G. (2017) "Seeds as ancestors, seeds as archives: Seed sovereignty and the politics of repatriation to native peoples," *American Indian Culture and Research Journal*, 41(3): 93–112.

Hodgson, G. M. (2015) *Conceptualizing Capitalism: Institutions, Evolution, Future*, Chicago: University of Chicago Press.

Holmén, H. (2005) "The state and agricultural intensification in sub-Saharan Africa," in *The African Food Crisis: Lessons from the Asian Green Revolution*, G. Djurfeldt, H. Holmén, M. Jirstrøm, and R. Larsson (eds.), pp. 87–112, Cambridge: CAB International.

Holodny, E. (2015) "The 13 fastest growing economies in the world," *World Economic Forum*, June 16.

Holt-Giménez, E. (2008) "Out of AGRA: The Green Revolution returns to Africa," *Development*, 51(4): 464–71.

Hountondji, P. (ed.) (1997a) *Endogenous Knowledge: Research Trails*, Dakar: CODESRIA.

Hountondji, P. (ed.) (1997b) "Introduction: Recentering Africa," in *Endogenous Knowledge: Research Trails*, Dakar: CODESRIA.

Houssou N., M. E. Johnson, S. Kolavalli, and C. Asante-Addo (2018) "Changes in Ghanaian farming systems: Stagnation or quiet transformation?," *Agriculture and Human Values*, 35(1): 41–66.

Hueler, H. (2014) "In Kenya, calls grow to lift controversial GMO ban," *Voice of America*, November 20, http://www.voanews.com/content/in-kenya-calls-grow-to-lift-controversial-gmo-ban/2527833.html, accessed July 3, 2015.

Huesing, J. (2018) "Fall armyworm in Africa: A guide for integrated pest management," presented at the SDSN carbon-free e-conference Responding to Fall Armyworm in Africa, October 22–26.

Huggins, C. (2017) *Agricultural Reform in Rwanda: Authoritarianism, Markets, and Zones of Governance*, London: Zed Books.

Humphrey, J., and O. Memedovic (2006) *Global Value Chain in the Agrifood Sector*, Vienna: UNIDO.

IFPRI (International Food Policy Research Institute) (2007a) "Assessing the potential impact of genetically modified crops in Ghana," http://programs.ifpri.org/pbs/pdf/pbsbriefghanaall.pdf, accessed May 1, 2011.

IFPRI (2007b) "Program for Biosafety Systems (PBS)—partners," http://programs.ifpri.org/pbs/pbspart.asp, accessed August 21, 2011.

IFPRI (2013) "Main findings: From subsistence to profit: Transforming smallholder farms," http://www.ifpri.org/publication/subsistence-profit?utm_source=New+At+IFPRI&utm_campaign=60e7f8b31b-New_at_IFPRI_August_26_2013&utm_medium=email&utm_term=0_7b974d57a5-60e7f8b31b-69105353, accessed July 7, 2015.

IFPRI (n.d.) "IFPRI Program for Biosafety Systems factsheet," http://www.cbd.int/doc/external/mop-04/ifpri-pbs-factsheet-ghana-en.pdf, accessed July 7, 2015.

Ignatov, A. (2017) "The earth as a gift-giving ancestor: Nietzsche's perspectivism and African animism," *Political Theory*, 45(1): 52–75.

Ignatova, J. A. (2015) "Seeds of Contestation: Genetically Modified Crops and the Politics of Agricultural Modernization," doctoral dissertation, College Park, University of Maryland.

Ignatova, J. A. (2017) "The 'philanthropic' gene: Biocapital and the new Green Revolution in Africa," *Third World Quarterly*, 38(10): 2258–75.

Illich, I. (2010) "Needs," in *The Development Dictionary: A Guide to Knowledge as Power*, W. Sachs (ed.), 2nd ed., pp. 95–110, London: Zed Books.

Independent Evaluation Group (2015) *World Bank Group Support to Public-Private Partnerships: Lessons from Experience in Client Countries, FY 02-12*, http://ieg. worldbankgroup.org/sites/default/files/Data/reports/chapters/ppp_eval_updated2.pdf, accessed December 21, 2020.

International Commission on the Future of Food and Agriculture (2006) *Manifesto on the Future of Seeds*, http://www.vandanashiva.org/wp-content/manifesto.pdf, accessed March 5, 2011.

ISAAA (2013) "Executive summary: Global status of commercialized biotech/GM crops: 2013," *ISAAA Brief 46-2013: Executive Summary*, http://www.isaaa.org, accessed January 17, 2016.

ISAAA (2017) "Global status of commercialized biotech/GM Crops in 2017: Biotech crop adoption surges as economic benefits accumulate in 22 years," *ISAAA Brief No. 53*, Ithaca, NY: ISAAA.

ISAAA (2018) "Executive summary: Global status of commercialized biotech/GM crops," *ISAAA Brief No. 54–2018*, http://www.isaaa.org/, accessed August 13, 2020.

ISAAA (2019) "Philippines approves Golden Rice for direct use as food and feed, or for processing," http://www.isaaa.org/kc/cropbiotechupdate/article/default.asp?ID=17900, accessed January 10, 2020.

ISAAA (n.d.a) "Biotech facts and trends 2014: Burkina Faso," https://www.isaaa.org/ resources/publications/biotech_country_facts_and_trends/download/Facts%20 and%20Trends%20-%20Burkina%20Faso.pdf, accessed July 5, 2015.

ISAAA (n.d.b) "Biotech facts and trends 2014: South Africa," https://www.isaaa.org/ resources/publications/biotech_country_facts_and_trends/download/Facts%20 and%20Trends%20-%20South%20Africa.pdf, accessed July 5, 2015.

ISAAA (n.d.c) "Nigeria approves confined field trial of cowpea," http://www.isaaa.org/kc/ cropbiotechupdate/article/default.asp?ID=4311, accessed July 5, 2015.

ISAAA (n.d.d) "Pocket K No. 16: Biotech crop highlights in 2018," https://www.isaaa.org/ resources/publications/pocketk/16/, accessed December 18, 2020.

James, C. (2014) "Global status of commercialized biotech/GM crops: 2014," *ISAAA Brief No. 49*, Ithaca, NY: ISAAA.

Jansen, K. (2003) "Crisis discourses and technology regulation in a weak state: Responses to a pesticide disaster in Honduras," *Development & Change*, 34(1): 45–66.

Jansen, K., and A. Gupta (2009) "Anticipating the future: 'biotechnology for the poor' as unrealized promise?," *Futures*, 41(7): 436–45.

Jasanoff, S. (2005) *Designs on Nature*, Princeton: Princeton University Press.

Jasanoff, S. (ed.) (2011) *Reframing Rights: Bioconstitutionalism in the Genetic Age*, Cambridge, MA: MIT Press.

Jasanoff, S. (2012) "Taking life: Private rights in public nature," in *Lively Capital*, K. Sunder Rajan (ed.), pp. 155–83, Durham: Duke University Press.

Jervens, M. (2014) "The political economy of agricultural statistics and input subsidies: Evidence from India, Nigeria, and Malawi," *Journal of Agrarian Change*, 14(1): 129–45.

Johnston, H. (2002) "Verification and proof in frame and discourse analysis," in *Methods of Social Movement Research*, B. Klandermans and S. Staggenborg (eds.), pp. 62–91, Minneapolis: University of Minnesota.

Juma, C. (2011) *The New Harvest: Agricultural Innovation in Africa*, Oxford: Oxford University Press.

Juma, C., and I. Serageldin (2007) "Freedom to innovate: Biotechnology in Africa's development," Report of the High-Level African Panel on Modern Biotechnology, Addis Ababa and Pretoria: African Union and New Partnership for Africa's Development (NEPAD).

Kaag, M., and A. Zoomers (eds.) (2014a) *The Global Land Grab: Behind the Hype*, London: Zed Books.

Kaag, M., and A. Zoomers (2014b) "Introduction: The global land grab hype—and why it is important to move beyond," in *The Global Land Grab: Behind the Hype*, M. Kaag and A. Zoomers (eds.), London: Zed Books.

Kahler, M., and D. Lake (eds.) (2003) *Governance in a Global Economy: Political Authority in Transition*, Princeton: Princeton University Press.

Kamola, I. (2019) *Making the World Global: U.S. Universities and the Production of the Global Imaginary*, Durham: Duke University Press.

Kansanga, M., P. Andersen, D. Kpienbaareh, S. Mason-Renton, K. Atuoye, Y. Sano, R. Antabe, and I. Luginaah (2018) "Traditional agriculture in transition: Examining the impacts of agricultural modernization on smallholder farming in Ghana under the new Green Revolution," *International Journal of Sustainable Development and World Ecology*, 26(1): 11–24.

Karungi, J., E. Adipala, S. Kyamanywa, M. W. Ogenga-Latigo, N. Oyobo, and L. E. N. Jackai (2000) "Pest management in cowpea part 2: Integrating planting time, plant density and insecticide application for management of cowpea field insect pests in eastern Uganda," *Crop Protection*, 19(4): 237–45.

Keck, M. E., and K. Sikkink (1998) *Activists beyond Borders: Advocacy Networks in International Politics*, Ithaca: Cornell University Press.

Kehoe, K. (2014) "Gates Foundation refutes report it fails African farmers," *Reuters*, November 5.

Kessides, I. N. (2004) *Reforming Infrastructure: Privatization, Regulation, and Competition*, Washington, DC: World Bank and Oxford University Press.

Khagram, S. (2004) *Dams and Development: Transnational Struggles for Water and Power*, Ithaca: Cornell University Press.

Khush, G. S. (2007) "Biotechnology: Public-private partnerships and intellectual property rights in the context of developing countries," in *Biodiversity and the Law: Biotechnology and the Law: Intellectual Property, Biotechnology, and Traditional Knowledge*, C. R. McManis (ed.), pp. 179–91, New York: Earthscan.

Kijima, Y., K. Otsuka, and D. Sserunkuuma (2011) "An inquiry into constraints on a Green Revolution in sub-Saharan Africa: The case of NERICA rice in Uganda," *World Development*, 39(1): 77–86.

Kinchy, A. (2012) *Seeds, Science, and Struggle: The Global Politics of Transgenic Crops*, Cambridge, MA: MIT Press.

Klein, N. (2000) *No Logo*, 2nd ed., New York: Picador.

Klein, N. (2007) *The Shock Doctrine: The Rise of Disaster Capitalism*, New York: Picador.

Kloppenburg, J. R. (2004) *First the Seed: The Political Economy of Plant Biotechnology*, 2nd ed., Madison: University of Wisconsin Press.

Kloppenburg, J. R. (2010) "Impending dispossession, enabling repossession: Biological open source and the recovery of seed sovereignty," *Journal of Agrarian Change*, 10(3): 367–88.

Kolavalli, S., K. Flaherty, R. Al-Hassan, and K. O. Baah (2010) "Do Comprehensive Africa Agriculture Development Program (CAADP) processes make a difference to country commitments to develop agriculture? The case of Ghana," *IFPRI Discussion Paper 01006*, Washington, DC: International Food Policy Research Institute.

Korten, D. C. (2001) *When Corporations Rule the World*, 2nd ed., Bloomfield, CT: Kumarian.

Kotschi, J. (2008) "Transgenic crops and their impact on biodiversity," *GAIA-Ecological Perspectives for Science and Society*, 17(1): 36–41.

Kuhlmann, K., and Y. Zhou (2016) "Seed policy harmonization in ECOWAS: The case of Ghana," *Working Paper*, https://www.syngentafoundation.org/sites/g/files/zhg576/f/seeds_policy_ghana_seed_case_study_jan16_0.pdf, accessed August 27, 2019.

Kumwenda, O. (2011) "GM on the rise in Africa," *Reuters*, March 31.

Kushwaha, S., A. S. Musa, J. Lowenberg-DeBoer, and J. Fulton. (2004) "Consumer acceptance of GMO cowpeas in sub-Saharan Africa," August 3, paper presented at the American Agricultural Economics Association Annual Meeting, Denver, CO.

Lambert, K. (2019) " 'It's all work and happiness on the farms': Agricultural development between the blocs in Nkrumah's Ghana," *Journal of African History*, 60(1): 25–44.

Leach, M., and R. Mearns (eds.) (1996) *The Lie of the Land: Challenging Received Wisdom on the African Environment*, London: International African Institute.

Li, T. M. (2007) *The Will to Improve: Governmentality, Development, and the Practice of Politics*, Durham: Duke University Press Books.

Li, T. M. (2011) "Centering labour in the land grab debate," *Journal of Peasant Studies*, 38(2): 281–98.

Liniger, H. P., R. Mekdaschi Studer, C. Hauert, and M. Gurtner (2011) *Sustainable Land Management in Practice—Guidelines and Best Practices for Sub-Saharan Africa*, TerrAfrica, Rome, Italy: World Overview of Conservation Approaches and Technologies (WOCAT) and Food and Agriculture Organization of the United Nations.

Lipschutz, R. D., and C. Fogel (2002) "Regulation for the rest of us? Global civil society and the privatization of transnational regulation," in *The Emergence of Private Authority in Global Governance*, R. B. Hall and T. Biersteker (eds.), pp. 115–40, Cambridge: Cambridge University Press.

Lipsky, M., and S. R. Smith (1989) "When social problems are treated as emergencies," *Social Science Review*, 63(1): 5–25.

Litfin, K. T. (1994) *Ozone Discourses: Science and Politics in Global Environmental Cooperation*, New York: Columbia University Press.

Little, P. D., and M. J. Watts (eds.) (1994) *Living under Contract: Contract Farming and Agrarian Transformation in Sub-Saharan Africa*, Madison: University of Wisconsin Press.

Losamills Consult Ltd. (2015) "The establishment of land banks," May 18 presentation for the National Lands Commission.

Losey, J. E., L. S. Raynor, and C. E. Carter (1999) "Transgenic pollen harms monarch larvae," *Nature*, 399(6733): 214.

Lukes, S. (2005) *Power: A Radical View*, 2nd ed., New York: Palgrave Macmillan.

Luna, J. K. (2019) "The chain of exploitation: Intersectional inequalities, capital accumulation, and resistance in Burkina Faso's cotton sector," *Journal of Peasant Studies*, 46(7): 1413–34.

Lynas, M. (2013) "The true story about who destroyed a genetically modified rice crop," *Slate*, August 26, http://www.slate.com/blogs/future_tense/2013/08/26/golden_rice_attack_in_philippines_anti_gmo_activists_lie_about_protest_and.html, accessed January 23, 2016.

Lynch, D., and D. Vogel (2001) *The Regulation of GMOs in Europe and the United States*, New York: Council on Foreign Relations.

Mamdani, M. (1996) *Citizen and Subject: Contemporary Africa and the Legacy of Late Colonialism*, Princeton: Princeton University Press.

Manji, A. (2006) *The Politics of Land Reform in Africa: From Communal Tenure to Free Markets*, London: Zed Books.

Manning, R. (2000) *Food's Frontier: The Next Green Revolution*, Berkeley: University of California Press.

Manning, R. (2005) *Against the Grain: How Agriculture Has Hijacked Civilization*, New York: North Point.

Manu, T. (2016) "Ghana trips over the TRIPS Agreement on Plant Breeders' Rights," *African Journal of Legal Studies*, 9: 20–45.

Marsden, K. (1990) *African Entrepreneurs: Pioneers of Development*, International Finance Corporation Discussion Paper, Washington, DC.

Martey et al. (2014) "Fertilizer adoption and use intensity among smallholder farmers in Northern Ghana: A case study of the AGRA soil health project," *Sustainable Agriculture Research*, 3(1): 24–36.

Martinez-Alier, J. (2002) *The Environmentalism of the Poor: A Study of Ecological Conflicts and Valuation*, Northampton, MA: Edward Elgar.

Marx, K. ([1867] 1976) *Capital, Volume 1*, B. Fowkes (trans.), New York: Penguin Classics.

Masters, W. A., J. Kuwornu, and D. Sarpong (2011) "Improving child nutrition through quality certification of infant foods: Scoping study from a randomized trial in Ghana," working paper (10/0828), London: International Growth Centre.

Matondi, P. B., K. Havnevik, and A. Beyene (2011) *Biofuels, Land Grabbing and Food Security in Africa*, London: Zed Books.

McAdam, D., S. Tarrow, and C. Tilly (2001) *Dynamics of Contention*, Cambridge: Cambridge University Press.

McCauley, J. F. (2003) "Plowing ahead: The effects of agricultural mechanization on land tenure in Burkina Faso," *Journal of Public and International Affairs*, 14: 1–27.

McGoey, L. (2014) "The philanthropic state: Market-state hybrids in the philanthrocapitalist turn," *Third World Quarterly*, 35(1): 109–25.

McGoey, L. (2015) *No Such Thing as a Free Gift: The Gates Foundation and the Price of Philanthropy*, London: Verso.

McGrath, M. (2013) "'Golden Rice' GM trial vandalized in the Philippines," *BBC News*, August 9, http://www.bbc.co.uk/news/science-environment-23632042, accessed July 3, 2015.

McIntyre, B. D., H. R. Herren, J. Wakhungu, and R. T. Watson (eds.) (2009) *Agriculture at a Crossroads: Global Report*, Washington, DC: IAASTD.

McKeon, N. (2009) *The United Nations and Civil Society: Legitimating Global Governance—Whose Voice?*, London: Zed Books.

McKeon, N. (2014) *The New Alliance for Food Security and Nutrition: A Coup for Corporate Capital?*, Amsterdam: Transnational Institute.

McKeon, N. (2015) *Food Security Governance: Empowering Communities, Regulating Corporations*, New York: Routledge.

McKeon, N. (2017) "Are equity and sustainability a likely outcome when foxes and hens share the same coop? Critiquing the concept of multistakeholder governance of food security," *Globalizations*, 14(3): 379–98.

McKeon, N., and P. McMichael (2014) "Land grabbing, investments in agriculture, and questions of governance," March 27, paper presented at the International Studies Association, Toronto, Canada.

McLean, S. A. (1987) "The Nairobi statement," The Enabling Environment Conference: Effective Private Sector Contribution to Development in Sub-Saharan Africa, Nairobi, Kenya, October 21–24. Geneva: Aga Kahn Foundation.

McManis, C. R. (ed.) (2007) *Biodiversity and the Law: Biotechnology and the Law: Intellectual Property, Biotechnology, and Traditional Knowledge*, New York: Earthscan.

McMichael, P. (ed.) (1994) *The Global Restructuring of Agro-Food Systems*, Ithaca: Cornell University Press.

McMichael, P. (1996) "Globalization: Myths and realities," *Rural Sociology*, 61(1): 25–52.

McMichael, P. (2000) "The power of food," *Agriculture and Human Values*, 17(1): 21–33.

McMichael, P. (2005) "Global development and the corporate food regime," *New Directions in the Sociology of Global Development*, 11: 269–303.

McMichael, P. (2009) "A food regime analysis of the 'world food crisis,'" *Agriculture and Human Values*, 26: 281–95.

McMichael, P. (2012) "The land grab and corporate food regime restructuring," *Journal of Peasant Studies*, 39(4): 681–701.

McMichael, P. (2013) "Value-chain agriculture and debt relations: Contradictory outcomes," *Third World Quarterly*, 34: 671–90.

Mgbeoji, I. (2006) *Global Biopiracy: Patents, Plants, and Indigenous Knowledge*, Ithaca: Cornell University Press.

Millar, D. (2014) "Endogenous development: Some issues of concern," *Development in Practice*, 24(5–6): 637–47.

Millenium Villages Project (n.d.) "About Millennium Villages sector strategy," http://mvs. millenniumvillages.org/about/sector-strategy/, accessed July 7, 2015.

Millstone, E., and T. Lang (2008) *The Atlas of Food: Who Eats What, Where, and Why*, Berkeley: University of California Press.

Ministry of Food and Agriculture Ghana (2017) "Planting for food and jobs," http://mofa. gov.gh/, accessed August 28, 2019.

Ministry of Food and Agriculture Ghana (n.d.) "Ghana Commercial Agriculture Project," https://mofa.gov.gh/site/projects/ghana-commercial-agriculture-project-gcap, accessed January 20, 2020.

Mitchell, T. (2002) *Rule of Experts: Egypt, Techno-Politics, Modernity*, Berkeley: University of California Press.

Mittal, A. (2009) "The blame game: Understanding structural causes of the food crisis," in *The Global Food Crisis: Governance Challenges and Opportunities*, J. Clapp and M. J. Cohen (eds.), pp. 13–28, Waterloo, Ontario: Centre for International Governance and Wilfrid Laurier University Press.

Mittal, A., and M. Moore (eds.) (2009) *Voices from Africa: African Farmers and Environmentalists Speak Out against a New Green Revolution in Africa*, Oakland, CA: Oakland Institute.

Mkandawire, T. (2007) "'Good governance': The itinerary of an idea," *Development in Practice*, 17(4/5): 679–81.

Mkandawire, T., and C. C. Soludo (1999) *Our Continent, Our Future: African Perspectives on Structural Adjustment*, Trenton: Africa World Press.

MOFA (Ministry of Food and Agriculture) (n.d.a) "Farmers' Day FAQs," *Republic of Ghana*, http://mofa.gov.gh/site/?page_id=6843, accessed July 2, 2015.

MOFA (n.d.b) "Update on AGRA programs and grants in Ghana," Republic of Ghana, http://mofa.gov.gh/site/?page_id=7588, accessed July 6, 2015.

Monsanto Company (2009) "Monsanto is on the verge of a technology explosion, executives tell investors at annual field event," *PR Newswire*, August 13, http://news.prnewswire.com/ViewContent.aspx?ACCT=109&STORY=/www/story/08-13-2009/0005076914, accessed February 12, 2012.

Monsanto Company (2013) "Monsanto Company 2013 annual report," http://www.monsanto.com/investors/documents/annual%20report/2013/monsanto-2013-annual-report.pdf, accessed July 6, 2015.

Monsanto Company (n.d.a) "Academic research agreements," http://www.monsanto.com/newsviews/pages/public-research-agreements.aspx, accessed January 23, 2016.

Monsanto Company (n.d.b) "Meet America's farmers," http://www.americasfarmers.com/?gclid=COzc6eSRuq4CFYmK4AodNjtYNQ, accessed January 23, 2016.

Monsanto Company (n.d.c) "Virus resistant cassava for Africa (VIRCA)," http://www.monsanto.com/improvingagriculture/pages/virus-resistant-cassava-for-africa.aspx, accessed July 5, 2015.

Monsanto Company (n.d.d) "Water efficient maize for Africa (WEMA)," http://www.monsanto.com/improvingagriculture/pages/water-efficient-maize-for-africa.aspx, accessed July 5, 2015.

Moola, S., and V. Munnik (2007) *GMOs in Africa: Food and Agriculture: Status Report 2007*, Melville, South Africa: African Centre for Biosafety.

Moore, P. (2014) "Has Greenpeace lost its moral compass?," http://www.allowgoldenricenow.org/moral-compass, accessed January 18, 2016.

Morris, M., V. A. Kelly, R. J. Kopicki, and D. Byerlee (2007) *Fertilizer Use in African Agriculture: Lessons Learned and Good Practice Guidelines*, Washington, DC: World Bank.

Morvaridi, B. (2012) "Capitalist philanthropy and hegemonic partnerships," *Third World Quarterly*, 33(7): 1191–210.

Moseley, W., M. Schnurr, and R. Bezner Kerr (2015) "Interrogating the technocratic (neoliberal) agenda for agricultural development and hunger alleviation in Africa," *African Geographical Review*, 34(1): 1–7.

Moseley, W., M. Schnurr, and R. Bezner Kerr (2016) *Africa's Green Revolution: Critical Perspectives on New Agricultural Technologies and Systems*, London: Routledge.

Moseley, W. G. (2017) "The new Green Revolution for Africa: A political ecology critique," *Brown Journal of World Affairs*, 23(2): 177–90.

Moseley, W. G., and L. C. Gray (2008) *Hanging by a Thread: Cotton, Globalization and Poverty in Africa*, Athens: Ohio University Press.

Mosse, D., S. Gupta, M. Mehta, V. Shah, J. F. Rees, and KRIBP Project Team (2002) "Brokered livelihoods: Debt, labour migration and development in tribal western Indian," *Journal of Development Studies*, 38(5); 59–88.

Mottiar, S., and M. Ngcoya (eds.) (2018) *Philanthropy in South Africa: Horizontality, Ubuntu and Social Justice*, Cape Town: HSRC.

Moyo, D. (2010) *Dead Aid: Why Aid Is Not Working and How There Is a Better Way for Africa*, New York: Farrar, Straus, and Giroux.

Müller, T. R. (2013) "The long shadow of Band Aid humanitarianism: Revisiting the dynamics between famine and celebrity," *Third World Quarterly*, 34(3): 470–84.

Muraguri, L. (2010) "Unplugged! An analysis of agricultural biotechnology PPPs in Kenya," *Journal of International Development*, 22(3): 289–307.

Mutua, M. W. (1995) "The Banjul Charter and the African cultural fingerprint," *Virginia Journal of International Law*, 35: 339–80.

MyJoyOnline (Ghana) (2014) "Angry farmers hit the streets over GMO," January 28, http://www.myjoyonline.com/news/2014/January-28th/angry-farmers-hit-the-streets-over-gmo.php, accessed July 3, 2015.

Nagoya Protocol on Access to Genetic Resources and the Fair and Equitable Sharing of Benefits Arising from their Utilization (2010), adopted by the Conference of the Parties to the Convention on Biological Diversity, October 29.

Nash/Zurich J. M. (2000) "This rice could save a million kids a year," *TIME magazine*, July 31.

National Geographic (2011) "Special series: 7 billion," http://ngm.nationalgeographic.com/2011/01/seven-billion/kunzig-text, accessed January 17, 2016.

Natural Justice and ABS (Access and Benefit-Sharing) Capacity Development Initiative (eds.) (n.d.) *Community Protocols in Africa: Lessons Learned for ABS Implementation*, Cape Town: Natural Justice and ABS Capacity Development Initiative.

Neeson, J. M. (1993) *Commoners: Common Rights, Enclosure and Social Change in England, 1700–1820*, Cambridge: Cambridge University Press.

Nestle, M. (2007) *Food Politics: How the Food Industry Influences Nutrition, and Health, Revised and Expanded Edition*, 2nd ed., Berkeley: University of California Press.

New Alliance for Food Security and Nutrition and Grow Africa (n.d.) *Joint Annual Progress Report: 2014–2015*, http://new-alliance.org/sites/default/files/resources/New%20Alliance%20Progress%20Report%202014-2015_0.pdf, accessed March 16, 2019.

New Partnership for Africa's Development (2018) *Country Overall Progress for Implementing the Malabo Declaration for Agriculture Transformation in Africa*, Midrand, South Africa: NEPAD.

New Partnership for Africa's Development (n.d.) "Comprehensive African Agricultural Development Programme (CAADP)," https://www.nepad.org/caadp, accessed July 16, 2020.

New Scientist (2004) "Monsanto failure," *New Scientist*, 181(2433): 7.

Ngcoya, M. (2009) "Ubuntu: Globalization, Accommodation, and Contestation in South Africa," doctoral dissertation, Washington, DC: American University.

Ngcoya, M., and N. Kumarakulasingham (2017) "The lived experience of food sovereignty: Indigenous crops and small-scale farming in Mtubatuba, South Africa," *Journal of Agrarian Change*, 17(3): 480–96.

Nin-Pratt, A., and L. McBride (2014) "Agricultural intensification in Ghana: Evaluating the optimist's case for a Green Revolution," *Food Policy*, 48: 153–67.

Nixon, R. (2011) *Slow Violence and the Environmentalism of the Poor*, Cambridge, MA: Harvard University Press.

Nkrumah, K. (1963) *Africa Must Unite*, New York: Frederick A. Praeger.

Nkrumah, K. (1964) *Consciencism: Philosophy and Ideology for De-Colonization*, New York: Monthly Review Press.

Nkrumah, K. ([1965] 1987) *Neocolonialism: The Last Stage of Neoimperialism*, Bedford: Panaf Books.

Nonzom, S., and G. Sumbali (2018) "Fall armyworm in Africa: Which 'race' is in the race, and why does it matter?" *Current Science*, 114(1): 27–8.

Nsafoah, A., M. Dicks, and C. Osei (2011) "Producers and consumer attitudes toward biotechnology in Ghana," February 5–8, paper presented at the Southern Agricultural Economics Association Annual Meeting, Corpus Christi, TX.

Nyantakyi-Frimpong, H., and R. Bezner Kerr (2015) "A political ecology of high-input agriculture in Northern Ghana," *African Geographical Review*, 34(1): 13–35.

Nyantakyi-Frimpong, H., and R. Bezner Kerr (2017) "Land grabbing, social differentiation, intensified migration and food security in northern Ghana," *Journal of Peasant Studies*, 44(2): 421–44.

Nyari, B. S. (2008) *Biofuel Landgrabbing in Northern Ghana*, http://www.tnrf.org/files/E-INFO-RAINS_Biofuel_land_grabbing_in_Northern_Ghana_Bakari_Nyari_2008.pdf, accessed July 7, 2015.

Oasa, E. K (1987) "The political economy of international agricultural research: A review of the CGIAR's response to criticisms of the 'Green Revolution,'" in *The Green Revolution Revisited: Critique and Alternatives*, B. Glaeser (ed.), pp. 13–55, London: Unwin Hyman.

Oliviera, G., and S. Hecht (2016) "Sacred groves, sacrifice zones and soy production: Globalization, intensification, and neo-nature in South America," *Journal of Peasant Studies*, 43(2): 257–85.

Organization of African Unity (1980) *Lagos Plan of Action for the Economic Development of Africa, 1980–2000*, Addis Ababa: African Union.

Osborne, D., and T. Gaebler (1993) *Reinventing Government: How the Entrepreneurial Spirit Is Transforming the Public Sector*, Reading, MA: Adison Wesley.

Ostrom, E. (1990) *Governing the Commons: The Evolution of Institutions for Collective Action*, Cambridge: Cambridge University Press.

Ostry, J. D., P. Loungani, and D. Furceri (2016) "Neoliberalism: Oversold?," *Finance and Development*, 53(2): 38–41.

Otieno, J. (2014) "Researchers in a fix over GMO ban," *East African*, August 16, http://www.theeastafrican.co.ke/news/Researchers-in-a-fix-over-GMO-ban/-/2558/2421314/-/item/0/-/qprcmnz/-/index.html, accessed July 5, 2015.

Ouma, S., M. Boeckler, and P. Lindner (2013) "Extending the margins of marketization: Frontier regions and the making of agro-export markets in northern Ghana," *Geoforum*, 48: 225–35.

Oya, C. (2012) "Contract farming in sub-Saharan Africa: A survey of approaches, debates and issues," *Journal of Agrarian Change*, 12: 1–33.

Paarlberg, R. (2001) *The Politics of Precaution: Genetically Modified Crops in Developing Countries*, Washington, DC: International Food Policy Research Institute.

Paarlberg, R. (2008) *Starved for Science: How Biotechnology Is Being Kept Out of Africa*, Cambridge: Harvard University Press.

Paarlberg, R. (2009) "The ethics of modern agriculture," *Society*, 46(1): 4–8.

Paddock, W. C. (1970) "How green is the Green Revolution?," *BioScience*, 20(16): 897–902.

Pambazuka News (2008) "AGRA, bio-piracy and food as social justice," *Pambazuka News*, October 4, http://pambazuka.org/en/category/comment/47258, accessed January 18, 2016.

Parbey, I. (2010) "Ghana: Biotechnology can change Ghana's agriculture," *Public Agenda* (Accra), August 30.

Parliament of Ghana (2020) "House Passes Plant Variety Protection Bill 2020 into Law after Adoption of Committee Report," https://www.parliament.gh/news?CO=97, November 5, accessed December 21, 2020.

Patel, R. (2012) *Stuffed and Starved: The Hidden Battle for the Food System*, 2nd ed., New York: Melville House.

Patel, R. (2013) "The long Green Revolution," *Journal of Peasant Studies*, 40(1): 1–63.

Patel, R., E. Holt-Giménez, and A. Shattuck (2009) "Ending Africa's hunger," *Nation*, September 21.

Patel, R., R. B. Kerr, L. Shumba, and L. Dakishoni (2015) "Cook, eat, man, woman: Understanding the New Alliance for Food Security and Nutrition, nutritionism and its alternatives in Malawi," *Journal of Peasant Studies*, 42(1): 21–44.

Pates, M. (2011) "Transforming Ghana's agriculture is focus of project," *Agweek*, June 6, http://www.agweek.com/event/article/id/18558/, accessed July 7, 2015.

Pechlaner, G. (2010) "The sociology of agriculture in transition: The political economy of agriculture after biotechnology," *Canadian Journal of Sociology*, 35(2): 243–69.

Peet, R., and M. Watts (eds.) (2004) *Liberation Ecologies: Environment, Development, Social Movements*, 2nd ed., London: Routledge.

Perbi, A. A. (2004) *A History of Indigenous Slavery in Ghana*, Accra: Sub-Saharan Publishers.

Perkins, J. H. (1997) *Geopolitics and the Green Revolution*, Oxford: Oxford University Press.

Peterson, V. S. (2003) *A Critical Rewriting of Global Political Economy: Integrating Reproductive, Productive and Virtual Economies*, London: Routledge.

Petrini, C. (2004) *Slow Food: The Case for Taste*, New York: Columbia University Press.

Pineau, C. (2005) *Africa: Open for Business*, 60 minutes.

Pingali, P. L. (2012) "Green Revolution: Impacts, limits, and the path ahead," *Proceedings of the National Academy of Sciences of the United States of America*, 109(31): 12302–8.

Pinstrup-Andersen, P., and E. Schiøler (2001) *Seeds of Contention: World Hunger and the Global Controversy over GM Crops*, Baltimore: Johns Hopkins University Press.

Pinstrup-Andersen, P., and P. B. R. Hazell (1985) "The impact of the Green Revolution and prospects for the future," *Food Reviews International*, 1(1): 1–25.

Plant Breeding E-Learning in Africa (n.d.) "Overview," https://pbea.agron.iastate.edu/about/overview, accessed August 14, 2020.

Polanyi, K. ([1944] 2001) *The Great Transformation: The Political and Economic Origins of Our Time*, 2nd ed., Boston: Beacon.

Prarie, M., ed. (2007) *Thomas Sankara Speaks: The Burkina Faso Revolution 1983–87*, 2nd ed., Atlanta: Pathfinder.

Prasanna, B. M., J. E. Huesing, R. Eddy, and V. M. Peschke (2018) *Fall Army Worm in Africa: A Guide for Integrated Pest Management*, 1st ed., Washington, DC: Feed the Future.

Praxis Africa (n.d.) "Broadcasting Ghana's development," http://www.praxissg.com/praxis-africa/the-farm-channel, accessed July 7, 2015.

Princen, T., M. F. Maniates, and K. Conca (eds.) (2002) *Confronting Consumption*, Cambridge, MA: MIT Press.

Puplampu, K. P., and G. O. Essegbey (2004) "Agricultural biotechnology and research in Ghana: Institutional capacities and policy options," *Perspectives on Global Development and Technology*, 3(3): 271–90.

Ragasa, C., I. Lambrecht, and D. S. Kufoalar (2018) "Limitations of contract farming as a pro-poor strategy: The case of maize outgrower schemes in Upper West Ghana," *World Development*, 102: 30–56.

Rahnema, M., and V. Bawtree (eds.) (2008) *The Post-Development Reader*, 2nd ed., London: Zed Books.

Rajagopal, B. (2003) *International Law from Below: Development, Social Movements and Third World Resistance*, Cambridge: Cambridge University Press.

Ramdas, K. N. (2011) "Philanthrocapitalism: Reflections on politics and policymaking," *Society*, 48(5): 393–6.

Raustiala, K., and D. G. Victor (2004) "The regime complex for plant genetic resources," *International Organization*, 58(2): 277–309.

Republic of Ghana (2013) *Plant Breeders Bill*, May 28, http://media.peacefmonline.com/docs/201312/919280493_445860.pdf, accessed July 6, 2015.

Republic of Ghana (2014) *Ghana Agriculture Sector Investment Programme (GASIP): Design Completion Report*, February 18, http://www.ifad.org/operations/projects/design/111/ghana.pdf, accessed July 6, 2015.

Reynaers, A., and G. De Graaf (2014) "Public values in public-private partnerships," *International Journal of Public Administration*, 37(2): 120–8.

Richards, P. (1985) *Indigenous Agricultural Revolution: Ecology and Food Production in West Africa*, Boulder: Westview.

Richards, P. (1993) "Cultivation: Knowledge or performance?," in *An Anthropological Critique of Development: The Growth of Ignorance*, M. Hoban (ed.), pp. 61–78, London: Routledge.

Richey, L. A., and S. Ponte (2011) *Brand Aid: Shopping Well to Save the World*, Minneapolis: University of Minnesota Press.

Rock, J. (2018) "We Are Not Starving: GMOs and Ghanaian Food Sovereignty Advocacy in the Age of the African Green Revolution," doctoral dissertation, Washington, DC: American University.

Rockefeller Foundation (n.d.a) "Impact investing," http://www.rockefellerfoundation.org/our-work/current-work/impact-investing, accessed July 1, 2015.

Rockefeller Foundation (n.d.b) "The world food problem, agriculture, and the Rockefeller Foundation," *100 Years: The Rockefeller Foundation*, http://rockefeller100.org/items/show/3780, accessed June 16, 2015.

Roe, E. M. (1995) "Except Africa: Postscript to a special section on development narratives," *World Development*, 23(9): 1065–9.

Roepstorff, T. M., and S. Wiggins (2011) "New global realities governing agribusiness," in *Agribusiness for Africa's prosperity*, K. K. Yumkella, P. M. Kormawa, T. M. Roepstorff, and A. M. Hawkins (eds.), Vienna: UNIDO.

Rogers, E. M. (2003) *Diffusion of Innovation*, 5th ed., New York: Free Press.

Rooney, K. (2019) "Why Jack Dorsey and other major tech figures are suddenly interested in Africa," *CNBC*, December 30.

Rose, N. (2001) "The politics of life itself," *Theory, Culture & Society*, 18(6): 1–30.

Rosi-Marshall, E. J., J. L. Tank, T. V. Royer, M. R. Whiles, M. Evans-White, C. Chambers, N. A. Griffiths, J. Pokelsek, and M. L. Stephen (2007) "Toxins in transgenic crop byproducts may affect headwater stream ecosystems," *Proceedings of the National Academy of Sciences*, 104(41): 16204–8.

Rudra, A. (1982) *Indian Agricultural Economics: Myths and Realities*, New Delhi: Allied.

Ruggie, J. G. (2004) "Reconstituting the global public domain—issues, actors, and practices," *European Journal of International Relations*, 10(4): 499–531.

Sachs, J. D. (2005) *The End of Poverty: Economic Possibilities for Our Time*, New York: Penguin.

Sachs, W. (ed.) (2010a) *The Development Dictionary: A Guide to Knowledge as Power*, 2nd ed., London: Zed Books.

Sachs, W. (ed.) (2010b) "Preface," in *The Development Dictionary: A Guide to Knowledge as Power*, 2nd ed., pp. vi–xiv, London: Zed Books.

SADA (Savannah Accelerated Development Authority) (2010) *SADA: Secretariat and Organizational Structure, Strategy and Work Plan 2010–2030*, Accra: Government of Ghana.

Sanjek, R. (ed.) (1990) *Fieldnotes: The Makings of Anthropology*, Ithaca, NY: Cornell University Press.

Schäferhoff, M., S. Campe, and C. Kaan (2003) "Transnational public-private partnerships in international relations: Making sense of concepts, research frameworks, and results," *International Studies Review*, 11(3): 451–74.

Schatz, E. (2009) *Political Ethnography: What Immersion Contributes to the Study of Power*, Chicago: University of Chicago Press.

Schumacher, E. F. ([1973] 2000) *Small Is Beautiful, 25th Anniversary Edition: Economics as If People Mattered: 25 Years Later … with Commentaries*, Vancouver: Hartley and Marks.

Schurman, R. (2017) "Building an alliance for biotechnology in Africa," *Journal of Agrarian Change*, 17(3): 441–58.

Schurman, R., and W. A. Munro (2010) *Fighting for the Future of Food: Activists versus Agribusiness in the Struggle of Biotechnology*, Minneapolis: University of Minnesota Press.

Scoones, I. (2005) "Contentious politics, contentious knowledges: Mobilising against GM crops in India, South Africa and Brazil," *IDS Working Paper 256*, Brighton: IDS.

Scoones, I. (2008) "Mobilizing against GM crops in India, South Africa, and Brazil," *Journal of Agrarian Change*, 8(2–3): 315–44.

Scoones, I. (2009) "Livelihoods perspectives and rural development," *Journal of Peasant Studies*, 36(1): 171–96.

Scoones, I. (2015) *Sustainable Livelihoods and Rural Development*, Black Point: Fernwood.

Scoones, I., and J. Thompson (eds.) (1994) *Beyond Farmer First: Rural People's Knowledge, Agricultural Research, and Extension Practice*, London: Intermediate Technology Publications.

Scoones, I., and J. Thompson (2011) "The politics of seed in Africa's Green Revolution: Alternative narratives and competing pathways," *IDS Bulletin*, 42(4): 1–23.

Scott, J. C. (1976) *The Moral Economy of the Peasant*, New Haven: Yale University Press.

Scott, J. C. (1987) *Weapons of the Weak: Everyday Forms of Peasant Resistance*, New Haven: Yale University Press.

Scott, J. C. (1998) *Seeing Like a State: How Certain Schemes to Improve the Human Condition Have Failed*, New Haven: Yale University Press.

Sell, S. K. (2003) *Private Power, Public Law: The Globalization of Intellectual Property Rights*, Cambridge: Cambridge University Press.

Sen, A. (1983) *Poverty and Famines: An Essay on Entitlement and Deprivation*, Oxford: Oxford University Press.

Sen, A. (1999) *Development as Freedom*, New York: Random House.

Sengooba, T., J. I. Cohen, and B. Zawedde (eds.) (2005) "Regulatory cooperation, using information, regional policies, and national expertise," proceedings of an East Africa Policy Roundtable, Entebbe, Uganda, April 18–20.

Serageldin, I. (1999) "Biotechnology and food security in the 21st century," *Science* 285(5426): 387–9.

Séralini, G. E., E. Clair, R. Mesnage, S. Gress, N. Defange, M. Malatesta, D. Hennequin, and J. S. de Vendômois (2012) "Long-term toxicity of a Roundup herbicide and a Roundup-tolerant genetically modified maize," *Food and Chemical Toxicology*, 50(11): 4221–31.

Séralini, G. E., E. Clair, R. Mesnage, S. Gress, N. Defange, M. Malatesta, D. Hennequin, and J. S. de Vendômois (2014) "Republished study: Long-term toxicity of a Roundup herbicide and a Roundup-tolerant genetically modified maize," *Environmental Sciences Europe*, 26(14): 1–17.

Sheeran, J. (2011) "Ending hunger now," *TED Talk*, July 28.

Shenggen, F., J. Brzeska, M. Keyzer, and A. Halsema (2013) *From Subsistence to Profit: Transforming Smallholder Farms*, Washington, DC: International Food Policy Research Institute.

Shepherd, A. (1981) "Agrarian change in northern Ghana: Public investment, capitalist farming and famine," in *Rural Development in Tropical Africa*, J. Heyer, P. Roberts, and G. Williams (eds.), London, UK: Palgrave Macmillan.

Shiva, V. (1991) *The Violence of the Green Revolution: Third World Agriculture, Ecology, Politics*, London: Zed Books.

Shiva, V. (1997) *Monocultures of the Mind: Perspectives on Biodiversity and Biotechnology*, London: Zed Books.

Shiva, V. (1999) *Biopiracy: The Plunder of Nature and Knowledge*, London: South End Press.

Shiva, V. (2000) *Tomorrow's Biodiversity*, New York: Thames & Hudson.

Shiva, V. (2001) *Protect or Plunder? Understanding Intellectual Property Rights*, London: Zed Books.

Shiva, V. (2009) "From seeds of suicide to seeds of hope: Why are Indian farmers committing suicide and how can we stop this strategy?," *Huffington Post*, April 28, http://www.huffingtonpost.com/vandana-shiva/from-seeds-of-suicide-to_b_192419.html, accessed January 17, 2016.

Shurtleff, W., and A. Aoyagi (2014) *History of Seventh-Day Adventist Work with Soyfoods, Vegetarianism, Meat Alternatives, Wheat Gluten, Dietary Fiber and Peanut Butter (1863–2013)*, Lafeyette: Soyinfo Center.

Silver, B., and S. S. Karatasi (2015) "Historical dynamics of capitalism and labor movements," in *Oxford Handbook of Social Movements*, D. Della Porta and M. Diani (eds.), pp. 1–11, Oxford: Oxford University Press.

Sisay, B. et al. (2018) "First report of the fall armyworm *Spodoptera frugiperda* (Lepidoptera: Noctuidae), natural enemies of Africa," *Journal of Applied Entomology*, 142(8): 800–4.

Smale, M., P. Zambrano, G. Gruère, J. Falck-Zepeda, I. Matuschke, D. Horna, L. Nagarajan, I. Yerramareddy, and H. Jones (2009) "Measuring the economic impacts of transgenic crops in developing agriculture during the first decade: Approaches, findings, and future directions," *IFPRI Food Policy Review No. 10*, Washington, DC: International Food Policy Research Institute.

Smith, E. (2012a) "Rice genomes: Making hybrid properties," in *Lively Capital: Biotechnologies, Ethics and Governance in Global Markets*, K. Sunder Rajan (ed.), pp. 184–210, Durham: Duke University Press.

Smith, J. M. (2012b) *Genetic Roulette: The Gamble of Our Lives*, 85 minutes.

Snow, D. A., Jr., E. B. Rochford, S. K. Worden, and R. D. Benford (1986) "Frame alignment processes, micromobilization, and movement participation," *American Sociological Review*, 51(4): 464–81.

Sobha, I. (2007) "Green Revolution: Impact on gender," *Journal of Human Ecology*, 22(2): 107–13.

Sorkin, A. R. (2014) "So Bill Gates has this idea for a history class …," *New York Times Magazine*, September 5.

Soybean Innovation Lab (n.d.) "Our mission," http://www.soybeaninnovationlab.illinois. edu/, accessed August 7, 2020.

Spann, M. (2017) "Politics of poverty: The post-2015 Sustainable Development Goals and the business of agriculture," *Globalizations*, 14(3): 360–78.

Spielman, D. J., F. Zaidi, and K. Flaherty (2011) "Changing donor priorities and strategies for agricultural R&D in developing countries: Evidence from Africa," *Working Paper 8* presented at the ASTI-IFPRI/FARA Conference, Accra, Ghana, December 5–7.

Sridhar, V. (2006) "Why do farmers commit suicide? The case of Andhra Pradesh," *Economic and Political Weekly*, 41(16): 1559–65.

Stiglitz, J. E., and S. J. Wallstein (1999) "Public-private technology partnerships: Promises and pitfalls," *American Behavioral Scientist*, 43(1): 52–73.

Stirling, A. (2010) "Keep it complex," *Nature*, 468: 1029–31.

Stone, G. D. (2007) "Agricultural deskilling and the spread of genetically modified cotton in Warangal," *Current Anthropology*, 48(1): 67–103.

Stone, G. D. (2002a) "Both sides now: Fallacies in the genetic-modification wars, implications for developing countries, and anthropological perspectives," *Current Anthropology*, 43(4): 611–30.

Stone, G. D. (2002b) "Commentary: Biotechnology and suicide in India," *Anthropology News*, 43(5): 5.

Stone, G. D. (2010) "The anthropology of GM crops," *Annual Review of Anthropology*, 39: 381–400.

Stone, G. D. (2015) "Golden Rice: Bringing a superfood down to earth," https:// fieldquestions.com/2015/08/28/golden-rice-bringing-a-superfood-down-to-earth/, accessed January 10, 2020.

Stone, G. D., and D. Glover (2017) "Disembedding grain: Golden Rice, the Green Revolution, and heirloom seeds in the Philippines," *Agriculture and Human Values*, 34(1): 87–102.

Strange, S. (1996) *The Retreat of the State: The Diffusion of Power in the World Economy*, Cambridge: Cambridge University Press.

Structural Adjustment Participatory Review Network (2004) *Structural Adjustment: The SAPRI Report: The Policy Roots of Economic Crisis, Poverty and Inequality*, London: Zed Books.

Stutz, B. (2010) "Companies put restrictions on research into GM crops," *Yale Environment 360*, May 13, http://e360.yale.edu/feature/companies_put_restrictions_ on_research_into_gm_crops/2273/, accessed January 23, 2016.

Sunder Rajan, K. (2006) *Biocapital: The Constitution of Postgenomic Life*, Durham: Duke University.

Sunder Rajan, K. (ed.) (2012) *Lively Capital: Biotechnologies, Ethics, and Governance in Global Markets*, Durham: Duke University Press.

Supreme Court of Canada (2004) *Monsanto Canada Inc. v Schmeiser*, 1 S.C.R. 902, 2004 SCC 34.

Syngenta (n.d.) "Grow more from less," http://www.syngenta.com/global/corporate/en/grow-more-from-less/Pages/grow-more-from-less.aspx, accessed January 17, 2016.

Tarrow, S. (2006) *The New Transnational Activism*, Cambridge: Cambridge University Press.

Taylor, M. (2015) *The Political Ecology of Climate Change Adaptation: Livelihoods, Agrarian Change and the Conflicts of Development*, London: Routledge.

Thompson, C. B. (2012) "Alliance for a Green Revolution in Africa (AGRA): Advancing theft of African genetic wealth," *Review of African Political Economy*, 39(132): 345–50.

Thompson, C. B. (2014) "Philanthrocapitalism: Appropriation of Africa's genetic wealth," *Review of African Political Economy*, 41(141): 389–405.

Thomson, J. A. (2002) *Genes for Africa: Genetically Modified Crops in the Developing World*, Landsdowne, South Africa: University of Cape Town Press.

Thorup, M. (2013) "Pro bono? On philanthrocapitalism as ideological answer to inequality," *Ephemera: Theory and Politics in Organization*, 13(3): 555–76.

Tilly, H. (2011) *Africa as a Living Laboratory: Empire, Development, and the Problem of Scientific Knowledge, 1870–1950*, Chicago: University of Chicago Press.

Tripp, R., and A. Mensah-Bonsu (2013) "Ghana's commercial seed sector: New incentives or continuing complacency?," *IFPRI Working Paper 32*, Washington, DC: International Food Policy Research Institute.

Tsing, A. L. (2005) *Friction: An Ethnography of Global Connection*, Princeton: Princeton University Press.

US Committee on Science House of Representatives (2003) *Plant Biotechnology Research and Development in Africa: Challenges and Opportunities, Hearing Before the Subcommittee on Research Committee on Science House of Representatives*, 108th Congress First Session, June 12.

US Congress, Office of Technology Assessment (1988) *New Developments in Biotechnology: U.S. Investment in Biotechnology-Special Report, OTA-BA-360*, Washington, DC: US Government Printing Office.

US Department of State (n.d.) "Agricultural policy," https://www.state.gov/agricultural-policy/, accessed January 14, 2020.

UNESOC (1991) *Appraisal and Review of the Impact of the Lagos Plan of Action on the Development and Expansion of Intra-African Trade*, 11th meeting of the conference of African ministers of trade, Addis Ababa, April 15–19.

United Nations General Assembly (1992) *Rio Declaration on Environment and Development*, Rio de Janiero, June 3–14.

UPOV (Union for the Protection of New Varieties of Plants) (1991) "International Convention for the Protection for the Protection of New Varieties of Plants of December 2, 1961, as Revised on November 10, 1972, on October 23, 1978, and on March 19, 1991," UPOV Convention, http://www.upov.int/en/publications/conventions/1991/act1991.htm, accessed July 6, 2015.

USAID (2012) "Ghana's private sector investment plan for agricultural development," http://pdf.usaid.gov/pdf_docs/PA00JZ75.pdf, accessed January 24, 2018.

USAID (2013) "Scaling Seeds and Technologies Partnership will accelerate progress to reduce hunger, poverty in Africa," *USAID Press Release*, June 28, http://www.usaid.gov/news-information/press-releases/scaling-seeds-and-technologies-partnership-will-accelerate-progress, accessed July 7, 2015.

USAID (n.d.) *Agricultural Biotechnology for Development*, Washington, DC: US Agency for International Development.

US Committee on Science House of Representatives (2003) "Plant Biotechnology Research and Development in Africa: Challenges and Opportunities," Hearing before the Subcommittee on Research Committee on Science House of Representatives, 108th Congress First Session, June 12.

Vandeman, A. M. (1995) "Management in a bottle: Pesticides and the deskilling of agriculture," *Review of Radical Political Economics*, 27(3): 49–55.

Van Lente, H. (2000) "Forceful futures: From promise to requirement," in *Contested Futures: A Sociology of Prospective Techno-Science*, N. Brown, B. Rappert, and A. Webster (eds.), , pp. 43–64, Aldershot: Ashgate.

Van Lente, H., C. Spitters, and A. Peine (2013) "Comparing technological hype cycles: Towards a theory," *Technological Forecasting and Social Change*, 80(8): 1615–28.

Via Campesina (2012) "International Conference of Peasants and Farmers: Stop land grabbing!," report and conclusions of the International Conference of Peasants and Farmers, Mali, November 17–19, 2011, http://viacampesina.org/downloads/pdf/en/mali-report-2012-en1.pdf, accessed July 2, 2015.

Via Campesina (n.d.) "What is La Via Campesina?," http://viacampesina.org/en/index.php/organisation-mainmenu-44/what-is-la-via-campesina-mainmenu-45, accessed July 3, 2015.

Vidal, J. (2010) "Why is the Gates Foundation investing in GM giant Monsanto?" *Guardian*, September 29.

WACCI (n.d.a) "Eligibility and selection criteria," http://wacci.ug.edu.gh/eligibility, accessed August 14, 2020.

WACCI (n.d.b) "Training the next generation of plant breeders in Africa for Africa," http://wacci.ug.edu.gh, accessed August 14, 2020.

Walker, S. (2001) "The TRIPS Agreement, sustainable development and the public interest," *IUCN Environmental Policy and Law*, 41: vii–60.

Waltz, E. (2009) "Battlefield," *Nature*, 461(3): 27–32.

Wambugu, F. (2001) "Taking the food out of our mouths," *Washington Post*, August 26.

Wambugu, F. (2015) "Ghana's biosafety law is model for Africa—Dr. Wambugu," *Graphic Online*, July 1, http://graphic.com.gh/business/business-news/45481-ghana-s-biosafety-law-is-model-for-africa-dr-wambugu.html, accessed July 3, 2015.

Wambugu, F., and D. Kamanga (eds.) (2014) *Biotechnology in Africa: Emergence, Initiatives and Future*, New York: Springer.

Webber, M. C., and P. Labaste (2010) *Building Competitiveness in Africa's Agriculture: A Guide to Value Chain Concepts and Applications*, Washington, DC: World Bank.

Wedeen, L. (2010) "Reflections on ethnographic work in political science," *Annual Review of Political Science*, 13: 255–72.

Weis, T. (2007) *The Global Food Economy: The Battle for the Future of Farming*, London: Zed Books.

Weissman, S. R. (1990) "Structural adjustment in Africa: Insights from the experiences of Ghana and Senegal," *World Development*, 18(12): 1621–34.

White, B., Jr., S. M. Borras, R. Hall, I. Scoones, and W. Wolford (2012) "The new enclosures: Critical perspectives on corporate land deals," *Journal of Peasant Studies*, 39(3–4): 619–47.

Wiemers, A. (2015) "A 'time of Agric': Rethinking the 'failure' of agricultural programs in 1970s Ghana," *World Development*, 66(C): 104–17.

Wild, S. (2017) "Invasive pest hits Africa," *Nature*, 543(March 2): 13–14.

Wilson, J. (2014) *Jeffrey Sachs: The Strange Case of Dr Shock and Mr Aid*, London: Verso.

World Bank (1981) *Accelerated Development in Sub-Saharan Africa: An Agenda for Action*, Washington, DC: World Bank Group.

World Bank (1987) *Chile Adjustment and Recovery*, report no. 6726-CH, Washington, DC: World Bank Group.

World Bank (1989a) *From Crisis to Sustainable growth—Sub Saharan Africa: A Long-Term Perspective Study (English)*, report no. 8209, November 30, Washington, DC: World Bank Group.

World Bank (1989b) *Ghana: Structural Adjustment for Growth*, report no. 7515-GH, January 23, Washington, DC: World Bank Group.

World Bank (1989c) *Developing the Private Sector: A Challenge for the World Bank Group*, Washington, DC: World Bank Group.

World Bank (1989d) *Infrastructure: The Foundation for Development: Outline for Ten-Year Plan of Action*, Working paper no. 21734, January, Washington, DC: World Bank Group.

World Bank (2007) *World Development Report 2008: Agriculture for Development*, Washington, DC: World Bank Group.

World Bank (2012a) *Africa Can Help Feed Africa: Removing Barriers to Regional Trade in Food Staples*, Washington, DC: World Bank.

World Bank (2012b) *Agribusiness Indicators: Ghana*, report no. 68163-GH, Washington, DC: World Bank.

World Bank (2012c) *Project Appraisal Document on a Proposed Credit in the Amount of SDR 64.5 Million (US$100 Million Equivalent) to the Republic of Ghana for a Commercial Agriculture Project*, February 27.

World Bank (2012d) "World Bank approves US$100 million for scaling up commercial agriculture in Ghana," *World Bank Press Release*, March 22, https://www.worldbank.org/en/news/press-release/2012/03/22/world-bank-approves-us100-million-for-scaling-up-commercial-agriculture-in-ghana, accessed August 13, 2020.

World Bank (2013) *Doing Business 2014: Understanding Regulations for Small and Medium-Size Enterprises*, 11th ed., Washington, DC: World Bank Group.

World Bank (2014) *Enabling the Business of Agriculture: 2015 Progress Report*, Washington, DC: World Bank Group.

World Bank (2018) "Creating an enabling environment (English)," *Responsible Agricultural Investment (RAI) Knowledge Into Action Note*, 5, Washington, DC: World Bank Group.

World Bank (n.d.a) "Data: Ghana," http://data.worldbank.org/country/ghana, accessed July 6, 2015.

World Bank (n.d.b) "Enabling the business of agriculture," https://eba.worldbank.org/en/data/exploreeconomies/ghana/2017, accessed August 7, 2020.

World Economic Forum (2014) "Africa's next billion," World Economic Forum Annual Meeting 2014, http://www.weforum.org/sessions/summary/africas-next-billion, accessed July 6, 2015.

WTO (World Trade Organization) (n.d.) "Uruguay round agreement: TRIPS," https://www.wto.org/english/docs_e/legal_e/27-trips_04c_e.htm, accessed July 6, 2015.

Wu, F., and W. Butz (2004) *The Future of Genetically Modified Crops: Lessons from the Green Revolution*, Santa Monica: Rand.

Yanow, D., and P. Schwartz-Shea (eds.) (2006) *Interpretation and Method: Empirical Research Methods and the Interpretive Turn*, New York: ME Sharpe.

Yapa, L. (1996a) "Improved seeds and constructed scarcity," in *Liberation Ecologies*, R. Peet and M. Watts (eds.), pp. 69–85, London: Routledge.

Yapa, L. (1996b) "What causes poverty?: A postmodern view," *Annals of the Association of American Geographers*, 86(4): 707–28.

Yaro, J. A. (2002) "The poor peasant: One label, different lives. The dynamics of rural livelihood strategies in the Gia-Kajelo community, Northern Ghana," *Norsk Geografisk Tidsskrift—Norwegian Journal of Geography*, 56: 10–20.

Yaro, J. A., and J. Hesselberg (2010) "The contours of poverty in northern Ghana: Policy implications for combating food insecurity," *Research Review*, 26(1): 81–112.

Yaro, J. A., J. K. Teye, and G. D. Torvikey (2017) "Agricultural commercialisation models, agrarian dynamics and local development in Ghana," *Journal of Peasant Studies*, 44(3): 538–54.

Yumkella, K. K., and P. M. Kormawa (2011) "Agribusiness for Africa's prosperity," lecture on behalf of UNIDO at the International Food Policy Research Institute, Washington, DC, October 25.

Yumkella, K. K., P. M. Kormawa, T. M. Roepstorff, and A. M. Hawkins (eds.) (2011) *Agribusiness for Africa's Prosperity*, Vienna: UNIDO.

Zerbe, N. (2004) "Feeding the famine," *Food Policy*, 29(6): 593–608.

Zerbe, N. (2005) *Agricultural Biotechnology Reconsidered: Western Narratives and African Alternatives*, Trenton, NJ: Africa World Press.

Zimmerer, K. S., and T. J. Bassett (eds.) (2003) *Political Ecology: An Integrative Approach to Geography and Environment-Development Studies*, New York: Guilford.

INDEX

www.ingramcontent.com/pod-product-compliance
Lightning Source LLC
Chambersburg PA
CBHW050440280326
41932CB00013BA/2183